The Smart Device

Johannes Väänänen

I0476679

Man Machine Interface (MMI), User Interface (UI), Human Interface. The most important aspect of any mobile device has different names in different cultures, but they all mean the same:

Linking the user – you – and a machine together.

The conversations in the book all come from the author's recollections, though they are not written to represent word-for-word transcripts. Rather, the author has retold them in a way that evokes the feeling and meaning what was said.

Edited by Jane Gossling

Cover design by Branko Balšić

Published by Vaka Väinämöinen OÜ

ISBN-13: 978-1523448524

ISBN-10: 1523448520

Prologue
Apple and The Man

Late spring 2002. The meeting room in Cupertino's Original Campus was packed with Apple's Human Interface designers – people who design the way Apple devices interact with users. They decide where to put the buttons and how a menu should pop-up.

This time these guys wouldn't have to do any designing. We were here to convince them, to sell to Apple the idea of a revolutionary user interface with a fully-functional prototype device.

Intel had seen an earlier version of it back in 2001, Telefonica had seen it, Motorola had seen it. Now was Apple's turn.

Page flicking was shown: as the finger was slid across the touch screen, pages would turn, just like a real book.

Auto-orientation was shown: turning the device portrait or landscape, the screen image rotated automatically too.

A full web browser was shown: in the days of the early, cramped mobile internet, the prototype was rendering normal web pages on a large virtual screen that could be zoomed and panned in real time.

A full, finger-operated QWERTY touch screen keyboard was shown and fast typing demonstrated using it.

The prototype was MyOrigo's **myDevice** - a new mobile phone from the distant wilderness of Finland. The Apple designers flipped the main menu pages back and forth, over and over again, wild-eyed with disbelief.

There was no knock on the door. The Man just stepped in, without any sort of fanfare.

Steve Jobs took the proto in his hands and began to swipe menu pages back and forth. The Man was hooked.

1.
The crafting of an innovator

I always wanted to be an inventor. Ever since I can remember, outlandish technical ideas would pop into my head and I would find peace from them only after I tried to make them real, to bring them into physical existence.

As a kid, implementing one's ideas is not easy. If the idea can possibly be built at all – which it most often cannot, of course – a kid does not have the tools, the knowledge, or the precision needed to actually make things work, to construct real functional machines.

The plan for implementation usually involves Lego, some electric motors from old toys, rubber bands and scotch tape. Every once in a while, one of those patched-together toy inventions actually does something interesting. Then a glimpse of true inventor's delight, the feeling of achievement that comes from deep within, is seen by the kid.

Most people grow out of that period, focusing on doing 'productive' things during their youth and adulthood. Like dating, accounting, hairdressing, piloting aircrafts and managing companies.

I never did.

As life went on, I got better at actually turning my inventions into reality. No more Lego: Commodore 64 would give me my first opportunity to see machines – albeit virtual machines created with software – that I had originally invented, actually working.

I was maybe 12 years old when I coded something on the good ol' C64 that nobody had done before, at least to my then-limited knowledge. That first taste of the joy that comes from seeing one's unique idea actually working is etched into my memory forever. That feeling beats everything else in this world. My fate was sealed - there would be no greater drug, no bigger motivation, than to invent, build and see the action of a working machine.

Fate would take me down the road to finally being able to 'create', to actually design and build physical devices that originated as ideas in my head. But I would have to wait for that time to come for more than 10 years, in the meantime trying to get my dose of 'invention-orgasms' from computer programming alone.

At times I thought it was a wait that would never end. I got increasingly proficient as a computer programmer and there was great joy in that too.

In my free time, when I was not chasing tail, I did what many do – repaired old cars, mainly Citroën's. It gave me some of the joy I thought physical inventions would. Maybe that was it, programming as a job, repairing cars as a hobby, balance of life. Building a family on the side, perhaps.

But my soul knew better. My dreams recoiled with impossible inventions, alien machines, waiting to become reality.

Finally, the door to real physical invention opened in 1999, almost by accident.

2.
Beginning of
Smart Phone era - first proto

Two men working on a prototype, precision tools and desktop computers with their side panels removed lying on the ground. The thing looked like a typical garage-made demo. A small display was attached to a bit of Styrofoam and some military spec-looking inertial modules, cable stack as thick as a thumb, ran from the display to the PC. The proto wasn't powered. Both men were trying to fit cable pack connectors to their awkward proper places, bending and stretching the cables as much as they could.

The room itself was remarkable, even under all the accumulated techno stuff. It resided in one of the oldest and finest apartment buildings in the city, boasting high, cathedral-like ceilings and intricate design features that made the ongoing work look out-of-place. Here, an unknown artist ought to be painting a future masterpiece or a once-famous author should be writing his final memoirs with a quill and ink.

The place belonged to J-P Metsävainio, an almost-famous Finnish architect back in the day. Weighing 140 kilograms, heavy too with design authority and endless social and marketing skills, he was plugging a VGA cable into the back of one of the desktops. All this, the techno-stuff around the room and on the tables, was a result of his ground-breaking concept – *when a display moves through the air, something should happen to it.*

That's where I came in. That other man was me. J-P had searched for somebody who could actually implement something concrete from his vision and found me, recommended by a software company I was working for at the time. I was 'the only guy who could maybe actually make sense of it and get it to work'.

It was summer of 1999 and I was 25 years old.

The first demo of the idea was to be simple, we decided. Imagine a big cardboard plate with a hole the size of credit card in the middle of it. Now imagine reading a newspaper by moving the cardboard around over the page – you can only see the area through the hole, but because of how the human brain works, you'd fairly soon have a grasp of what the whole page looked like in your mind. You could even read text quite comfortably, since paragraphs have limited width, meaning the entire width of a paragraph would be visible through the hole.

And that was it. Motion-controlled browsing of material larger than the display itself! Finally, we could bring full-sized internet pages to the small screens of mobile phones. Actual cardboard demos were made, but now something more substantial was needed to make the idea real. A real display and real motion control had to be brought together.

The building of this demo, the prototype, was no trivial task back then. But I thought I saw a way. Parts and components were ordered, and finally, after a couple of months, implementation started at J-P's office.

The only computers powerful enough to run such a demo were desktop PCs and the only sensors able to provide motion information for such purposes were expensive motion capture sensors used by the movie folks in Hollywood. I decided to make the proto by attaching a 7 inch VGA display (a hard item to come by in itself back then) and an Ascension Bird motion sensor to a Styrofoam board, and hooking them up via long cables to the most powerful Windows NT desktop PC we could find at a reasonable price. The display driver card in the PC had to be top notch too, as we also had some ideas about presenting some motion sensitive 3D-material on the screen. The Ascension sensor included an external reference magnetic field pulse receiver, a one litre box, that needed to be in close proximity to the motion sensor attached to the display. In all, the prototype was quite a mess from a hardware point of view, but it worked as planned.

The software part of the proto was easy for me once the development environment for the Ascension motion sensor was set up. I was a seasoned Windows programmer and knew user interface programming and 3D graphics inside out, so that wasn't going to be a problem. A couple of weeks later, the proto had a software packet built with an extensive mixture of C++ and J++.

Building the prototype took three months, with me designing the hardware architecture and implementing all the software and J-P providing the graphics elements and helping with assembling the hardware. Co-operation was surprisingly easy, for the first time in my life I felt I had actually met a Fellow Innovator, someone on my wavelength.

The proto turned out beautifully. While it looked physically quite garage-made, the motion control was super smooth. It was convincing proof that true internet

---

could really work on a small screen, even on a mobile phone.

Holding the Styrofoam plate in one hand and moving it through the air, the embedded display panned to a different part of the full-sized web page in real time. It was an amazing illusion; the brain could almost immediately create a mental 'picture' of the whole page. It also had the ability to zoom; moving the device nearer or further away zoomed the web page accordingly, again, in real time.

We couldn't stop playing with it and everybody who saw the proto was awestruck. This was the era when mobile screens were black and white, 200x120 pixel, small patches lurking above the phone buttons. Mobile internet was called WAP, showing 3-5 compressed lines of simple text. And here we were, demonstrating real-time panning and zooming of a complete web page, on a mobile screen...

We were sure we were onto a winner. This would revolutionize the mobile phone market and, in the process, make us as rich as the Rockefellers.

Now the next task, surely an easy one after building the proto, was to sell the invention to a company big enough to actually put it into 100 million mobile phones! Surely the proto would make all the potential buyers see the value of such technology...

3.
'The Devil that escaped Nokia' and Microcell

While I was busy building the motion control prototype, still unaware of the real commercial potential of our invention, the mobile industry in Finland was exploding. Big changes lurked on the horizon.

Nokia was by far the biggest mobile phone manufacturer in the world, employing directly or indirectly about half of Finland's engineers. It had surpassed Siemens, Motorola and Ericsson years ago. Nokia was considered the best place to work back then, people standing in line to get interviewed and hired. Think of the status of Google as an employer today and you get the picture. Many socially-talented ground-level employees reached middle management after just a few years on the career treadmill and Nokia granted generous stock options to anybody who was somebody on the company ladder. Managers were getting rich left, right and centre. The engineers, the actual technical can-do people, were not.

Nokia didn't have the slightest intention of compensating their engineers better. Why should they? As it was, they paid the highest salaries when it came to 'ordinary' engineers, and job opportunities in Finland that didn't involve Nokia in one way or another were scarce.

How long could such a company culture last? Maybe decades, if no game changer came along... But nature finds its equilibrium and it seems this even applies to the ecosystems in large companies like Nokia too. Often a single event, a single individual can upset a seemingly everlasting status quo.

Meet 'the one that got away', the stone in Nokia's gigantic shoe: Jyrki Hallikainen, former senior VP at Nokia, a man possessing such a charisma that it reached evangelical levels when he spoke to his colleagues and team members. Such a man – a possible game changer and pretender to the throne - was always feared by traditional company managers who jealously guarded their own positions to the bitter end. Nokia was no exception, and a rift began growing between Jyrki and his bosses. Finally, Jyrki had fled Nokia and was now recruiting Nokia's top design talent into his new company, Microcell.

Jyrki understood that offering higher salaries and stock options for Nokia's best technical engineers would get him the people that Nokia relied on – the people not properly compensated by their organization. That, plus his unbelievable appeal and the ability to talk almost anybody into almost anything. If ever there were a Jesus preaching to mobile engineers, it was Jyrki. He knew just what to say and what to offer.

Engineers in Finland started to become aware of a new employer, a game changer, in Oulu. Oulu – being the true 'design lab' for Nokia and once the mobile development centre for the whole of Europe, was small enough that everybody soon knew somebody who had gone to Microcell. Big salaries and shares, was the story. Rumours were even circulating about a 'signing-on fee', a new concept in Finland for ground level employees. Once you signed your job contract, you immediately got 20.000 Euros in your bank account! This was unheard of, and tantalizing as hell to bread-and-butter engineers.

Where did the money to finance Microcell come from, given that, after all, this kind of recruiting required serious bucks? For a social genius like Jyrki, it proved easy to talk second runner-up in the mobile business, Ericsson, into ordering and partly pre-paying a complete mobile phone from his then-miniscule company. Of course it helped ever so slightly that Ericsson would get to sell the design of the best (now former) Nokia engineers – touché!

Some new component vendors also wanted to get into the booming mobile business. Since big players like Nokia were not willing to risk designing new, unknown components into their mass-produced phones, it was hard to penetrate the component markets. Jyrki capitalized on that too – National Instrument were so eager to get into the game that they financed Jyrki's company with millions of dollars based solely on mere promises that their components would be utilized in Microcell's designs. Once again, Jyrki's people-skills sealed the deal.

And so it went. Within a year, Microcell had been able to recruit almost 100 of Nokia's best designers. Many of the recruits came from companies doing subcontracting to Nokia, even more from Nokia directly. Very few engineers were even interviewed from outside those circles, but as usual, there were a couple of dark horses that were pulled in by luck or accident.

The engineers who jumped ship were offered a finder's fee, monetary compensation, if they could recommend somebody talented enough to hire who did not yet figure on the 'possible recruits' list. Mechanical engineer Timo Remes, a former Nokia employee who had just signed a contract with Jyrki, thought he knew of one. He recommended to Jyrki that if there was something new and innovative to be done in Microcell's mobile business, there was one guy the company had to have on board.

Thus I was recruited.

The new millennium had just started and J-P's and my revolutionary prototype was up and running. But our initial enthusiasm about selling the idea to the obvious customer, Nokia, had faded with the dawning realisation of what NIH means: Not Invented Here. Nokia's managers had no interest whatsoever in some prototype presented by 'outsiders', as surely they could do better. It didn't even matter what we showed them, it could have been a working time machine, the effect and response would have been the same. A ridiculous offer was made to us along the lines of 'we'll pay you 20.000 Euros and you'll give us all the rights to the idea, along with a gag clause'. So that was the end of that with Nokia.

So I was free to do whatever I wanted then. J-P and I didn't know anybody in the mobile business outside Finland, and back then going and presenting your invention in the US, for example, was too complicated, too expensive and too 'different' to the business culture of Finland, we thought.

The basic idea, the concept of motion control in a mobile device, nevertheless still appealed to me. I didn't want to forget about it and treat it like just another of the demo projects that come and go. Also, now knowing the level of sensor and mobile technologies available, my head was buzzing with new inventions that I could surely turn into a reality – if there was just some way of finding a mobile phone company that would let me build them.

Big companies such as Nokia would never let me do what I wanted, I understood that very clearly now. I would be crammed into a cubicle, tasked with fine tuning over and over again some obscure piece of SIM card managing code, and forgotten along with the rest of the grey masses of engineers doing the same routine work, day in day out. I was no great salesman, my social skills were poor by any standard, so I couldn't talk my way up the all-powerful company ladder. Even then, pure technical skills in a company like Nokia had almost no value if you couldn't market yourself to get promoted.

Microcell, as offered to me by my cousin Timo, was a different kettle of fish. It would be a place that would listen to my inventions and innovations. Jyrki Hallikainen, 'the Evangelist of our times', would recognize my talent and let me change the mobile world. Well, according to Timo's sales pitch, anyway.

Trusting my cousin, I reasoned that getting into Microcell and then showing the proto to Jyrki would get me working on it in no time at all. Maybe I could push my inventions to the mobile world after all? So, off to the interview!

Two months passed. I was going to see Jyrki for the first time, after the initial job interview with Jyrki's right hand engineer, Sakke Pelkonen a few weeks earlier. It was clear that I was already hired based on that interview, now it was just a matter of me being happy with the proposed compensation. It was a lot higher than my salary in my previous jobs, but I had to see the boss, 'the Evangelist', before I could commit.

Jyrki greeted me in his casual, nice-to-see-a-distant-friend style. We met in his office, big enough to hold both a meeting table and several whiteboards. Dirty coffee cups lay all over the place, people ran in and out all the time, peering round the door and asking the boss things. In this chaotic environment, Jyrki was completely relaxed and unnerved. After a couple of minutes of feather-light everyday chat about my background and my family ties, I knew I wanted to work for this man. We didn't even discuss any technical aspects of my future work or my position within the Microcell hierarchy. I was sure he'd let me become who I really was deep within, an inventor with the capability to change the world. That was the Jyrki Effect, he was able to make everybody feel what they needed to feel. We didn't even have to discuss numbers, I accepted everything the way Jyrki proposed it. A couple of signatures later, I was in.

I was hypnotized. But unlike other hypnoses, this didn't wear off until much, much later.

4.
From Microcell to 'my' MyOrigo

How do all those new tech start-up companies, new seeds that can blossom into a great big money-growing trees someday, get born? It seems there are three common ways it can happen.

Some companies start 'in the garage' from scratch without funding, building a business via small subcontracting deals at first, living almost hand-to-mouth. Oftentimes the founders have a 'day job' elsewhere to keep up with their living costs, making the first years of running such a company basically a very time-consuming hobby. If they are disciplined enough, they can slowly gather enough money and knowledge to develop their own technology to some level of maturity, later building their own products even.

Others get a nice lump sum of funding to start with, so-called 'seed money', from either the entrepreneurs themselves (which means they had to be rich or connected to begin with), or from real investors who see potential in the business idea. Sometimes, if the founders are connected enough, the funding comes in a form of binding project order from some big company ... and with that binding order some banks loan money. No matter how the funding was acquired, these companies can proceed immediately to develop their own stuff, spending money to do it the way they want to, not having to worry about getting in small ad-hoc projects just to keep running.

The third kind is the most complex one. Management Buy-Out, or MBO: some key personnel group from a bigger company takes off, establishes a new company, and acquires some rights (yes, those boring patents) or technologies from that parent company. They pay that bigger company either in form of money, delayed royalty payments, or shares from their own new business – often all of these. This way, a technology that the big company might never have used to its full potential, gets developed by a smaller, more dedicated unit that just focuses on that piece. Everybody wins – ex-employees get their own company and rich up if they succeed, the big firm gets rid of tech they don't want to develop further, and they get paid for it. Well, in theory, anyway.

As it seems to be with my life here on this earth, the establishment of my first 'own' company was not as simple as any single one of those ways. No, it was rather a strange mixture of all of them, all tangled up with Microcell and Jyrki.

Hypnotized or not, my employment at Microcell started with chaos. With a high-speed company like that chaos was to be expected by anybody who knew the industry. I didn't back then, so for me the general working methods and project management styles of mobile phone companies were kind of a shock. I quickly learned to swim along, writing 'spaghetti code and sloppy direct interfaces' to fulfil each week's goals only to trash the whole module a week later, when new features would be added. It was like being in demo creation mode all the time, but Microcell was actually implementing a real mass-production phone for Ericsson...

I saw no way to actually make any of my own ideas function with this kind of working style. And as a mere software engineer, I had no power to even try to put any unspecified new features anywhere. It would mean an immediate kick in the nuts from higher ranking fellow designers and middle-managers. The only 'novelty' I could muster was a WYSIWYG simulator running on a PC where that phone's software code, applications and user interface could be tested without the actual phone hardware. To the mobile engineers it was something new, even Jyrki admired it. They could see the perfect look-alike phone function on their computer screens without the actual physical phone, pushing the phone buttons on the screen and all that! Back then it was new... but I felt no joy doing minor stuff like that.

It was time to make something happen, then. If not, I would be coding that poo forever with that grey mass of engineers.

I arranged a demo session with Jyrki and Microcell, where J-P and I would show our prototype – and hopefully get some kind of traction inside Microcell to start doing something with it. After all, Jyrki had made me feel like I would be granted the space to innovate and invent, now was the time to test the reality behind my expectations.

The demo session went as expected. Many of my fellow engineers tried to keep straight, unimpressed faces when real-time panning and zooming with motion control was shown – after all I was a junior in their mobile circles, so no kudos would be appropriate. I got that undertone and let J-P pretty much do the whole presentation. After all it was only a sales pitch to get me working on something interesting, maybe also to get J-P on board Microcell.

Jyrki and some of his higher ranking buddies were very impressed, however, and they were not afraid to show it. Right after the session, general planning was started to make something "real" out of the demo.

Now it must be remembered that Jyrki was not only a self-righteous businessman and the unchallenged leader of his company, but also a social mastermind. He obviously understood instantly that developing anything so outlandish and new would be impossible with Microcell's current organization and style. There were too many ex-Nokia mobile industry experts, too many know-alls, who knew exactly how a phone must be made, and, more importantly, how it must not be made. In practice, blocking most new ideas, with all too good technical reasons.

To circumvent the situation, a decision was made: Jyrki, J-P and I would found a new company, officially independent of Microcell, to cook up a ground-breaking new mobile phone. One that would revolutionize the market. And earn billions, not millions, for us. One with motion control, and who knows what else... The demo was just a teasing glimpse of what I had in mind.

That moment in time, that decision, finally ended my old software-engineer-dreaming-of-inventions life. The new life, a true inventor's life with all its perks, began. An entrepreneurship crash course from A to Z!

So, six months after meeting Jyrki, I had my new company that would focus only on building my inventions, loosely based in J-P's and my demo device. I called it 'my' company as I personally owned most shares – but the final control was in Jyrki's hands, as the principal investor and chairman of the board, he had over 50% through his investing companies... JP was also in, not as a private shareholder like me, but with an ownership level equal to mine through his company.

Microcell and Jyrki were strongly backing the new company, of course, both moneywise and resource-wise. But regardless of that I felt my hands were not tied in any way. It was a real start-up company, named MyOrigo after the motion control general idea. Each user has a physical *origo*, a zero point in front of the face, around which the motion control happens. The My-first syllable could mean the user's 'my', or maybe just mine, I thought silently after inventing the name.

The journey of a start-up is seldom without rain clouds, especially when founded like this. This was a little bit like an MBO from Microcell's viewpoint, since the idea was presented to them with the original motivation to utilize it there.

It was also a bit like garage-started one, since by virtue of the configuration of the companies with no direct ownership from Microcell, MyOrigo would have to do subcontracting to Microcell to get money for its own actual design and development work.

And it was a bit like a seed-money start-up, because Godfather Jyrki arranged a good enough deal with Microcell in the beginning that subcontracting carried the company forward quite well, even leaving some time to work on its own technology.

Microcell's personnel were understandably not always happy about these developments, especially later when it became clear we were succeeding in building something more revolutionary than they. But that unhappiness was never really manifested as bad co-operation between the two companies, more on a personal level with some of the employees. My radical design ideas, cockiness and tendency to look past ex-Nokia know-alls did not help with this – from their viewpoint a junior SW guy, me, was suddenly running his own company and building his own inventions, with full support from the Evangelist, Jyrki. The keys to the kingdom were in my hands. It was just out of place, too much too fast, many surely thought.

It was September 2000, and I was the first employee and manager of MyOrigo Inc... I was officially an inventor entrepreneur now. *Prinssi Eversti.*

5.
Building MyOrigo

MyOrigo's original strategy, laid down by Jyrki, was faithful to the Microcell business model: 'Own Design and Manufacturing', a term Jyrki had presented to the world a year ago. ODM was to make the leading mobile companies like Nokia, Ericsson or Motorola into just brands, the true design and even manufacturing being done by smaller and more agile companies like Microcell.

ODM was supposed to be good for small tech companies, because it removed the expensive sales and marketing tasks from their shoulders. Brands like Ericsson would handle the marketing with their endless budgets. ODM could focus on what it did best, design and manufacturing, just like the name implies. Of course ODM never gained the awareness or praise from the actual customers, the people buying their products branded by those large, well-known companies. But who cared, as long as the business was lucrative. In those days there were no billion-dollar sales of small start-ups to big dogs like Google nowadays, so that recognition was not so valuable back then. There was more value in building real stuff and getting your money from that.

MyOrigo was to be an ODM too. With one critical difference: MyOrigo would not design phones based solely on big companies' specifications. It would present a ready-made concept of a revolutionary phone and just add some customer-specific features (Little did I know it did not mean just a logo on the front ...).

That made it easy then, I thought. No need to build a massive sales and marketing organization, I just hire the best technical guys I can find, make the perfect phone with a new kind of user interface, sell it to be branded by a big company (preferably Ericsson who was already Microcell's customer) and get as rich as an average Rothschild.

Before any engineers, the first thing I needed was somebody to handle the daily chores, you know, banking, paying taxes, buying computers, payroll and so on. In short, that somebody would run the company while I was doing my thing, inventing. I wanted to have somebody with an intricate knowledge of running a small company, but at the same time, somebody I knew would not question my decisions. I did not want to get into a fight over the cost of my previous day's business lunch in the middle of all-important inventing work!

There was only one such person I knew. My friend since childhood, a former Tar Tsar from my home village, more than 20 years older than me. He was called Tapsa, and he was all business. An unsurpassed ladies man, he had run (and bankrupted) numerous companies of his own, and was quite the storyteller. A Citroen-man to the heart, which explains my earlier hobby of repairing old Citroens... Everybody either liked or hated him, there was no middle ground. I thought he would also act as a filter, if somebody did not like him, that somebody would not be suitable to work in my company anyway.

So Tapsa was in. As a bonus, he had a son who was perfect to head the IT department of MyOrigo (freeing me from tweaking the company's constantly misbehaving email servers).

That took care of the basic infrastructure of my company.

Then, the actual workforce. I planned to implement all I could myself to get it just right, but many additional resources would be needed anyway. I took on a few of the software guys I had originally recruited to Microcell, basically employing them at MyOrigo as 'old employees'. I knew I could rely on those guys, all the more so because they also had PC software backgrounds so they were more open to implementing mobile devices in a new way.

Those first software guys formed the hard core of MyOrigo's software team, managing singlehandedly the code closest to the hardware to the end of MyOrigo, and beyond... Upi Pietikäinen, Janne Rosberg, Eero Nurkkala. The System Software team.

Hardware designers? I didn't know any.

Mechanics? Only one I knew was my cousin Timo, a mechanical maestro by any standard, but he was too busy at Microcell. They would not let him go. And like they say, never mix family and business...

So I made it my first mission to find somebody external to design the hardware for MyOrigo's still totally unspecified prototypes, and rely on Timo supporting my company's mechanical design in his free time, so to speak. I was sure I could gather resources over time, when things moved forward (and I was right, a couple of years later MyOrigo had more than 100 employees).

To get money into the newly-founded MyOrigo, me and the software guys started doing subcontracting work for

Microcell, just as planned. So basically, the guys did a full day's work of their old job, I was doing less than half a day if that, and the precious other half of each day was left to building up MyOrigo to be a real company, not just a name.

The subcontracting money flowing in from Microcell was just enough to keep us running. Any more would have been a waste anyway, I did not have a plan yet as to what to do with my newly-vested powers. The first proto was a long way off from being a mass producible device, I'd have to re-invent and rationalize the original ideas to make them actually doable with a real life mobile phone – only then could the real work at MyOrigo start.

6.
An inventor's dream

What makes an inventor tick?

Getting others to cheer for one's ideas and the resulting devices? Being named 'Inventor of the year'? No.

Getting tons of riches, funnelled to sports cars, private jets and Ukrainian ballerina girlfriends? Rap music video style? Nope.

Hopes of making the world a better place? No, except for the most naive of us... The balance of good and evil must and will always remain. Every new successful invention will be most likely used for both, anyway. Inventing does not really promote either one of those human traits in the long term, new inventions just open the door for more of them – extra dosage of good *and* evil.

What can it be then? Why bother inventing at all in this ready-made world? Wouldn't it be easier just to be a 'normal' businessman, selling ecologically grown potatoes by the ton?

It's that lovin' feeling. The deep self-awareness of having successfully done something nobody did before, ever. I call it *Inventor's delight.* I believe every child at some point gets to experience that godly emotion in one circumstance or another. In a sense, that feeling is the total opposite of emotions resulting from social interaction. It's a feeling of self-admiration that must originate from a mammal chemical hormone level, from the oldest realms of what being a human means. We have the capability to bend physical reality to suit our will – we can be the true masters of our universe.

I believe that originally we inventors are the ones that made it all happen, in the first place. Not politicians, not businessmen, not opinion leaders – they only act on a recoil of an original invention. Also, we are the ones that can change the game in most unexpected ways. We built the current civilization, and we can also make it change. This means that to the Establishment, whoever they are at any given time, 'we the inventors' pose the greatest risks. Their comfortable earning models and hyper-elevated position in society's hierarchy can be turned around in the blink of an eye if some crazy inventor comes up with... free energy... free information (ah..., that's already been done, it's called the internet) ... possibility to travel at great speeds without roads or government control... possibility to build a new home at miniscule cost in a single day, anywhere... Risks, great and dangerous risks to the Establishment. And the damnest thing is most of us don't give a dime about 'changing the world' - we are just looking to get our *Inventor's delight* dose! Think Tesla and you get the picture.

Now it was my time to hit it big.

I had a vision – a dream - of a totally new kind of mobile device, now I would materialize it with my new company. Backed by the best resources I could find, funded by Jyrki and Microcell, I was sure I could do it.

The general idea of viewing web pages and documents

much larger than the physical screen of a mobile phone, and panning and zooming the view with motion control, was beautiful – it was the focus point of our first demo. I believed it held the key to the future of MyOrigo. After all, people were going to want to view the same web pages they saw on their pc's display on their mobiles too.

Under the hood, at the core of the system, was a technology called *virtual screen*. It was, for a lack of a better word, fooling the application software like a web browser to think that they had a large screen to render to – like a pc-sized monitor or even bigger. Of course physically that 'large monitor' did not exist, it was only a virtual one inside the device's memory buffers, but it did not matter to the applications. The important thing was the user never had to wait for applications to 'finish' rendering before the virtual screen could be seen – no, they could view and pan around the virtual screen all the time, just like they would 'look' at a normal pc display. That's what separated the virtual screen approach from a normal 'application that renders the part the user wants to see'-system. If virtual screen is omitted, and the web browser application (for example) is given the task of rendering a part of the web page user wants to see, there are bound to be smaller or larger delays whenever user wants to see a different part of the web page. The user would have to wait for the application to render the part of the web page they wanted to see – just like scrolling a complex web page with a slow computer, every scroll notch results in a wait for processing before the page looks right again. And any delays amount to a poor user experience, everybody knows that. Virtual screen eliminates that delay – the whole viewable virtual screen area is all the time available to pan and zoom, without any delays, if properly implemented.

Technically the virtual screen approach was quite

straightforward to code. There was just a need to create a so-called off-screen RAM buffer, which in this case would be larger than the actual physical screen, and force all the application-level rendering operations to that buffer. The web browser app could take all the time it needed to download and render the data to the virtual screen, the user could all that time be panning around the virtual screen and even click a button on it before the complete web page had downloaded, for example.

But how could the user control the virtual screen, pan and zoom around it? In our demo device it was that famous 'hole-in-the-cardboard'-trick – but could it work on an actual mobile phone, with the technology available in the year 2000?

After some serious thinking I came to the conclusion that mostly all the user interface ideas presented in our first demo device were poorly suited for a hand-held device – especially the original motion control method, moving the device around in air.

That surely ingenious motion control would be cool if only... well, you'd always have to have a meter of empty space around you and you wouldn't mind moving and holding the device in the air, in arbitrary locations in the air to be more precise, while you used a web browser. From the everyday user's perspective, the required grandiose gestures in the air might look cool, but it would be quite impractical in any real life scenario. Waving your device around in the air on a bus or a train might not hurt anybody as long as you were careful, but it surely wouldn't be comfortable or 'cool'.

On top of the usability issues, there were also technical limitations to implement our original motion control's 'hole in the cardboard' - method. There was no reliable method to determine the exact to-the-millimetre-precise location of a wireless device in an arbitrary three-dimensional space – actually there still isn't in 2015.

Determining the exact location of an electronic device on the near surface of our planet Earth is no piece of cake. Nowadays GPS, Glonass and Galileo satellites can provide location accuracy of up to half a meter – that's no good for user interface purposes, where even millimetre scale movements have to be taken into account to provide a smooth and instant user experience, in motion control terms. Accelerometers and gyros can only iterate positions in space, which makes them prone to integral errors. A small inaccuracy of a sensor reading of a couple seconds before starts to accumulate and form an ever-increasing error that soon proves fatal to any user-experience concentrated motion detection system. Many newbies spend their time in trying to perform accelerometer based micro-navigation (the term I use to signify millimetre precision positional information in arbitrary 3D-space) work, first compensating for the bias errors of accelerometers, later trying to use fuzzy logic to determine whether or not the device is actually moving in relation to Earth's surface (or the user). They will fail. The Coriolis effect alone makes their efforts useless, even if theoretically 'perfect sensors' were available – and today's sensors are far from perfect, even more so in the beginning of 2000's. Millimetre precision micro-navigation in free air just seems impossible without some external reference beacon, which was of course impossible to provide for a mobile phone that would be used in all kinds of locations.

Years later in the second half of the 2000's, gaming companies hit the same wall, and finally came up with their own reference-based solutions. They could do so because those gaming consoles are used statically in living rooms, so external reference points could be placed in the room. Sony PlayStation 3's motion

control technique uses a video camera on top of
the TV set to provide a reference position to
the game controllers which have a bulb of light
in the end of then to make the video camera
'see' them. The Microsoft Xbox uses a
statically located video camera and infrared
environment digitizer to follow the user's
motions in real space. The Nintendo Wii does
not even attempt to provide micro navigational
capabilities, it just detects 'swings' and
other motion events like that, and provides a
pointer functionality via an (again external)
reference infrared sensor located on top of the
TV.

Other prominent features of the demo, such as flicking
through pre-downloaded search results of a web browser
like browsing a stack of papers, were at the time too heavy
and network bandwidth-intensive to be practically
implemented, I already understood that.

So, apart from the concept of the virtual screen, almost
everything in our original demo had to go. I'd just have to
follow the original mind-set, the general WOW feeling of
the demo device, and re-invent all the actual user interface
methods to suit a real world hand-held mobile phone. It
was Tabula Rasa time.

First things first: Motion control, the big new theme that
we had brought to the scene, the actual sales point of our
demo, had to be rediscovered. Rediscovered in a form that
would not require large movements or holding the phone
in any certain 'location' for too long which would become
a strain to the user's hand.

I had fiddled around with a tilt-and-scroll idea already in
the first proto – tilting the device would scroll the image
on screen towards the direction of the rotation. Like a
drop of water starts to slide across a tray when you tilt it.
It never really felt pleasurable to use, something intuitive

was lacking – tilt & scroll was about as comfortable as scrolling the web browser screen with physical up-down-left-right buttons. Well, the only good points of tilt & scroll in motion control terms were that it required little space for movement since the device did not have to moved, just tilted – and it would be possible to implement with just accelerometer sensors, the only motion sensor type really commercially available in those times. But other than that, tilt & scroll was a loser – it never created any illusion of hand/eye-coordination or naturalness of usage.

I had all the knowledge and all the pieces needed to invent a new motion control system that would work in a mobile phone, and would provide a WOW-effect with its great usability. My brain just had not pieced it all together yet, and any conscious effort to come up with a solution did not yield better results that that mundane tilt & scroll system. I felt I had to give myself the space and the time to come up with the solution through intuition.

Now how does one invent? Can inventing be a logical process, where small steps of deduction lead to the final outcome, a new invention? At least for me it never worked like that. My inventions pop up from the black depths of my subconscious, following the all-bright light of intuition. And they usually surface at the most unexpected moments. Maybe that moment would come...

In the meantime, there was much to do in other areas. For the first time, I got to really think what a complete mobile phone actually has to do. For example, one must be able to place a call, more precisely, to dial a number... Since I wanted to maximize the area for the display, the only real options were to place physical keys on the back of the device accompanied by a second smaller display, or to make the main display sensitive to touch. Keys on the back of the device would kind of defeat the whole purpose of simple usability - so it must be the touch panel then: humans want to use fingers to dial their phone numbers, so I thought let's provide them with that. I postponed the actual selection of touch screen technology for later and went with the assumption 'it can be done' - there were PDAs on the market with touch panels already, albeit they all required the use of plastic pens (styluses).

The menu system was an interesting nut. Back in the day all the mobile phones had complex – even cryptic – menu systems that required users to click through sprawling tree structures to find the function they needed – whether it be opening calendar application or setting a ring tone. Nokia's menu layout was considered the best in the market – and to my taste, even that sucked. I thought of using the original demo's 'page stack' idea for a menu – the whole device's content and functions would be organized into a stack of pages, like a book, that would always remain in same order, on a single level. So finding a menu function would be as simple as browsing a physical book to find a page you are looking for. Again, the tedious implementation details could wait, as well as the selection of the *page order* for that kind of 'electronic book'. Anyway, it would 'only' be software level work that would be fully specified and coded much later.
I was still a tad unsure about the actual physical appearance of the future device. All the PDAs were like square blocks, much wider than phones – and practically all the phones were equipped with tiny displays on top of physical keyboard, form factors designed to be narrow enough that they were comfortable to be held and used with one hand. Maybe a 'best of both worlds' -approach would be suitable for my future mobile phone – a narrow phone-like device with a large, but similarly narrow screen? No definite decision about this could be made before the actual availability of displays and technologies regarding touch screens was known. Deciding the shape of

the device - its form factor in the industry terms - would have to wait.

All the conceptual pieces were by now coming together quite well in my head. Just one thing – the most important thing, actually – was missing: how to control the virtual screen with motion control? With what kind of physical movements would the user pan and zoom around a web page or document larger than the physical screen?

It came to me on a shivering cold, windy September evening. Visiting my girlfriend-of-the-hour, I was using a small hand-held vanity mirror to check on an annoying pimple on my face. A sudden realization popped to my head like a cork on top of wavy waters: I am panning my face with hand mirror, using motion control – tilting – to select the part I want to see! 'This is totally intuitive and requires only a tilt to work well', I thought. It was also immediately clear that a simple 2-axis accelerometer could provide the needed orientation data of the device. That moment, I had invented a totally new method of motion controlling virtual screen – a form of digital hand mirror, so to speak. 'Flash of Genius', I thought in my narcissistic inner dialogue; using this method would be totally natural for anybody who'd ever looked at their own face in a hand mirror – and everybody has...

Thus *Virtual Mirror*™ was born. The big riddle was now solved; I was sure I could implement this idea easily enough on a mobile phone.

As it usually is with me, the Inventor's Delight-moment did not happen at that time, the time of birth of the theoretical idea. I'd have to wait until the first prototype version of Virtual Mirror would actually operate in front of my eyes, only then would I feel the moment of ultimate satisfaction filling my veins. The emotion I was feeling now was more like a relief, as there would be no conceptual pieces missing from then on.

I talked J-P into understanding and accepting the idea of Virtual Mirror. We figured together that the idea would be – anyway – quite difficult to convincingly explain to a layman without a working proto, so its *Grande premiere* would have to wait. And also, the patenting process should be started (as I knew already from studying into what 'patenting' actually was, for a few months) before really opening up the Virtual Mirror idea to any 3rd parties. No problem – in the meantime we would present our first demo to investors and recruits, nobody would have to know just yet that our final motion control solution for mobile phones would actually be quite different indeed.

7.
Touchy Feely HaptiTouch™ - the original idea of a finger-usable touch screen

The concept of our future mobile phone – *a smart phone* - was now quite well formed in my head, inventing the Virtual Mirror had been the final stroke of a conceptually-sized paintbrush. It was time to delve in the nagging details of how to actually build it all, and refine the concepts from there on. That motion-control-mirror-trick would be easy, even for first protos, but the touch panel... how to actually source one in the year 2000? It was a critical piece of tech still missing.

At a concept level I knew we needed a touch screen, and all my previous work had been based on the assumption that I could find or build a suitable touch screen system. Now, before the actual prototyping of mobile phone to-be could start, I had to produce the real deal.

There were some touch sensitive technologies available, but practically all of them were designed to be used by stylus, a pen-like object made out of plastic. That would not do for my taste. I wanted easy usability for a normal finger - which I believed was the key to a new user experience that would replace the conventional click-a-button-keyboard.

Some experimenting with available stylus-driven devices like Alcatel One Touch Com and many Windows CE- and Palm-based PDAs revealed that the 'technology in trend', resistive touch panel, was not the way to go. Only barely useable with a stylus, the resistive panels totally failed under finger control providing sludge feedback and very poor accuracy – not to mention the hindering the screen light throughput with its multiple layers of plastic on top of it. And, with resistive touch panel tech, there would be no curves, no extending the touch panel outside the display area, whereas I wanted the whole front panel of the device to be touch sensitive.

But hey, today everybody and their grandmother - including iPhone - uses capacitive touch screens, why didn't we go that route back in 2000? Well, because that technology did not exist in any useable form back then! The finger-friendly touch screen was science fiction...

It seemed inevitable that I had to invent the basic touch panel technology myself. And I had an idea ready – why not use the force distribution from well-known static mechanical force distribution equations to determine the point of origin of the force – in this case, a finger?

I figured it was a pretty neat original idea – any surface could be rigged to act as a touch panel. From a practical viewpoint, suitable micro-scale force sensors had just entered the market, mainly targeting robotics applications – so electrical components were readily available, at least so it seemed. And I had already quite a good grasp of the physics behind statistical force and weight distribution – in theory. I made some calls, ordered some components, and started to build a first working prototype of the idea on my fake mahogany desk.

I acquired some Honeywell Force Sensors, FSS series, and hooked them up to a National Instruments 24-bit A/D -converter box connected to my desktop PC via an RS-232 serial cable. A simple transparent plastic plate, fixed loosely with screws on top of an aluminium cradle holding 3 force sensors, in a triangular configuration. A one-day software environment setup, and a half day of intensive Visual C++ 6.0 coding resulted in a surprisingly accurate estimate of the place of origin of the force on the surface. The force could be applied with any object, finger, tongue, pen, even paintbrush - this first prototype provided an accurate location info of the force with about one Newton force - or 100 grams' weight, in a horizontal configuration. A triangular configuration of the force sensors resulted in interesting mathematics. Force applied outside the sensor triangle yielded the opposite force sensor to actually *show less force or weight*, in effect 'lifting weight' from those force sensors. The math worked well, but naturally required significant pre-tension to be applied to each force sensor - otherwise the 'lifting force' at the sensor could not be measured at all and the touch location would become corrupted outside the triangle. I was confident that the suitable pre-tension could be provided for a mass production version of the technology, and preferred my 3 sensor triangle configuration more than the possible (more expensive) 4 sensor square formation.

a 4 sensor configuration would practically never experience 'lifting force' but would have other problems, mainly aligning each sensor with micrometre precision to the right height so that all of them would have the same effect from the touch panel. Electronic configuration of the system was simple enough, I figured it could work on a mass produced device. Sensors consumed only some milliwatts of energy when in use, and a 16-bit standard A/D-converter would do just fine in measuring the signal - my 24-bit configuration was a necessary overkill to see the limitations in sensitivity of the sensors themselves.

A week of tedious but interesting mechanical crafting, coding and debugging ensued. Finally – and rather suddenly, after one critical bug was fixed in the code - the prototype worked well enough that I felt the surge of *Inventor's delight* – one of those moments I live for. It was a rainy, misty October afternoon, but I was on mental sunshine. I did a small tap-dance around my table in total silence, smirking with extreme self-satisfaction. Narcissism? Sure. Egoism? No doubt. Satisfactory? To die for. A new, revolutionary touch panel technology had surfaced by my mortal hand!

Now I could confidently demonstrate the prototype to engineers – and to J-P and Jyrki. Placing a finger on top of phone-sized transparent plastic board resulted in circle on my desktop PC's screen – and the circle was drawn at the relative position of the touch point, accurately enough that the WOW-effect was guaranteed. After all, it was the first time that a 'dead' plastic plate, without any integrated electronics, could sense the point of touch, and the accuracy was convincing! It was also self-evident that the miniscule force sensors could be integrated to almost any portable device's electronics with ease.

Second WOW-effect was the child of the actual physical mechanism behind the system– the circle on screen grew in size as the force that finger was pressed against the panel increased. This was not only touch sensitive, but also force sensitive touch panel! The endless application possibilities – capital letters from soft keyboard by simply pressing harder, a painting application with the capability to draw heavier strokes with heavier finger pressure, touch screen sensitivity that could be adjusted – by software – from feather-sensitive to military-grade tough touch force.

Only later when MyOrigo started patenting this system of

force-sensitive touch did I learn that since 1970's there had been attempts to fabricate touch screens like this. Using force distribution on the surface to determine the point of touch, was not as original an idea as I initially thought. Well, good that I did not know about those failed trials beforehand, I might have given up before even trying. And I would have never had *that* Inventor's delight – moment...

J-P had been my support all this time, playing the invisible hand as the board member and user interface advisor (practically, advising and discussing with me during the prototyping). Now, seeing the force driven touch screen in action, he seemed to gain more and more momentum to his own feel as to whether this kind mixture of things that I was proposing could be done. Virtual Mirror was still only talk, but this novel touch screen technology actually seemed to work in practice too!

The last element of our touch technology surfaced in discussions with J-P. We decided, mainly due to J-P's conviction about the necessity of such a thing, that the final device would include tactile feedback. A small physical 'tick', a hint of vibration, would notify the user of successful keypress on the touch screen. J-P knew well from his previous studies that the latency of touch sensation is orders of magnitude smaller than hearing or seeing – so this kind of tactile or *haptic* feedback would be a superior experience compared to the 'beep'-sounds and visual feedback that could be given for keypresses.

After all this, this technology's proper trademarked name was not hard to come by:

HaptiTouch.

8.
The first-generation prototypes (2000-2001)

All concept work must end sometime, to give room to prototyping. Ideas follow dreams, concepts follow ideas, and prototypes must follow concepts. Otherwise all is in vain, and inventor shrinks to a mere dreamer who never could bring his 'visions' to reality.

Virtual screen, Virtual Mirror, HaptiTouch... lots of nice names for inventions, concepts, that did not really exist yet in consumer space physical reality. Well, a proto was made of HaptiTouch, but it was a mere tech trial to validate the basic method of operation, not really capable of showing how it would work in an end product.

The practical concept of prototyping is simple: gather as much ready-made stuff as you can, glue it together, and add as little of your own stuff as possible. This way you can do it fast, with low cost, and if you are good – with results that can lead toward building the next prototypes with more of your own design, the next ones, and after that, maybe a mass product too...

As we were about to produce a hand-held device with the highest possible processing power (to be sure there was enough juice to operate the virtual screen), the most logical thing was to look for a *reference design kit* built around the existing processors of the time. The most powerful mobile processor at year 2000 was Intel's StrongArm SA-1110, an Arm-based processor with good-sized internal caches and 32-bit memory bus. Conveniently, Intel provided just the reference design kit needed, including a PDA format (4:3) display, two sandwich-sized circuit boards, and a battery. It could be packed into a plastic encasing roughly 15 x 10 x 8 cm in measurements, I calculated. Hardly mobile phone-sized, but small enough to be considered a hand-held device – and more importantly, flexible enough by design that the extra hardware components needed for Virtual Mirror and HaptiTouch could be integrated using small additional circuit boards.

The only thing I lacked, really, was a hardware designer who could put it all together after designing the extra boards Virtual Mirror and HaptiTouch required. Microcell had its share of HW guys but none of them had ever designed much around high powered CPU's and big displays – they were traditional mobile phone architecture engineers, squeezing last drop of performance or battery life out of 8-bit processors and small monochrome displays. So I had to look elsewhere.

A suggestion came from one senior project manager at Microcell: I was supposed to contact Juhani Putkinen, a seasoned HW designer as well as a prototype builder himself. The project manager's reference was reassuring: "If you need to make something work, and you don't care about social relations or empty chit-chat talk, Juhani is the guy who will do it, no matter what.".

That was the short interlude to guide me to meet one of the most influential persons regarding my later life.

I met Juhani at Microcell's Kuopio office – a small town 300 kilometres away from Oulu and the headquarters. A short, yankee-haired spry older fellow greeted me with a powerful handshake and military gaze – I mumbled my name in reply and wondered if that man would be compatible with my ideas of the forthcoming hardware development. Probably not – but maybe he was so proficient, that I could make a compromise and make myself compatible with his working style? These were the thoughts that ran through my head from the first minute of our meeting.

The more we discussed, the more obvious it became to me that I had met a very capable hardware designer – and I liked the forthright attitude which meant no time had to be spent on niceties. All dialogue was straight, not compensated to better suit 'the customer expectations', or whatever the half-lying bullshit talking normal sales guys do is called. Juhani Putkinen was a man of his word, was my first embodiment of this guy.

Juhani told me quite a lot about his history. He had designed and built electronic devices for the military. He had been designing stuff on the actual factory lines, at a component level, so he understood also what mass production meant in practical sense. And most importantly, he had experience of high-speed memory buses and 16/32-bit processors, which was what I needed.

The think I liked about Juhani, and about Finnish engineers in general, is that they concentrate on the problematic parts of any given project. An American engineer starts to speak about things that already work and problems ('challenges') they have successfully solved; a Finnish engineer begins to talk about unsolved problems first. And that I like – why talk about things already done, it's a waste of time. Leave that to the marketing guys!

Engineers should always concentrate on the unsolved problems, not pride themselves on solved ones and rest in their laurels, admiring their accomplishments. Juhani was obviously on the same page with me on that: when we discussed the prototype I was planning, he found mainly negative things to say about the hardware configuration I proposed. And it was not critique for the sake of critique, it was clever and accurate observations and notes about the possible shortcomings of my planned HW. It was straight talk and well justified!

I took the bait and hired Juhani on the spot. He would be a subcontractor, working under his own company, but in practice acting as MyOrigo employee whenever the situation needed it – like when talking to the component vendors or customers. Although I thought it might be best not to let Juhani participate in customer meetings, as his straight-talking ways might scare the most 'civilized-country-grown-up'-customers away.

The meeting progressed into the late hours of the evening, and we were still talking tech. The format of the MyOrigo first phone prototype started to develop:

- Intel's StrongArm reference design kit would be the basis, heart of the design
- Extra circuit boards would be built for an accelerometer (for Virtual Mirror) and each force sensor (for HaptiTouch), and Juhani would attach them to proper interfaces on that reference kit's boards
- All the software would be coded with MyOrigo existing system software team – they could build on top of the bootloader and other ready-made stuff provided by Intel with the reference design kit

---

- Mechanics would be designed and manufactured by Microcell mechanics engineers – the same ones that had helped me with the mechanical construction of the HaptiTouch tech proof proto just a month earlier
- Actual GSM phone functionality, or even the HW required to make a call, would not be provided with this first proto – the main goal would be to test and validate the user interface concepts and the operation of HaptiTouch in a real wireless device.

From then on it was all quite straightforward. The accelerometer for motion sensing was selected from Analog Devices' collection, the force sensor was to be the same one used in my tech proto – the Honeywell FSS series sensor, four for each prototype. Two prototype units in total were to be built, one for debugging and programming, the other to be enclosed in mechanics for testing - and showing.

Three months after our initial meeting with Juhani, I had a working prototype in my hand, enclosed in a dull gun metal-grey mechanics, display covered with the barely transparent PCB plastic plate that HaptiTouch required. It was called MOP2, short for MyOrigo Prototype 2 (prototype 1 being the original demo device, albeit it was actually constructed before MyOrigo came into existence).

MOP2's Virtual Mirror demo software worked on the level that made me feel a dose of Inventor's Delight entering my veins – using the device to pan around a web page bigger than the physical screen got me high – it was just like I imagined it to be. It was intuitive, fast, precise and... cool. Yes, Juhani had been able to deliver!

Scrupulous testing of HaptiTouch on MOP2 was not all happy times, though. The touch screen worked on some levels, but inaccurately and jerkily. It seemed the mechanics were fighting to keep the changing touch forces from being measured continuously and accurately. It was obvious that implementing flawlessly working HaptiTouch to a mass product would not be a walk in the park – MOP2 was far from it, for sure.

Even with its shortcoming in the touch screen department, MOP2 convinced me – and quite soon also J-P and Jyrki – that this was the way to go. It proved the processing power and sensor technology was available to really make a mass product around our UI concepts – and it proved we could design it, especially now that Juhani was on the team. It was time for the next step.

While toying with MOP2, there was an actual business issue at hand: MyOrigo was about to hire its first 'real' CEO, finally leaving me to do what I did best, inventing and technical design as CTO. Harri Vatanen, the ex-Sonera SmartTrust boss, was up for an interview. Jyrki had done a good job of explaining the background story to Harri prior to our meeting: MyOrigo was building the next revolution in the mobile industry, and funding was supposedly secured by Jyrki's contacts and authority. Jyrki wanted J-P and me to lure Harri in with our original demo device, that massive pile of hardware that pre-dated MyOrigo – and I understood why. The original demo was still the most convincing item we had to show for – especially the 'hole-in-the-cardboard' -motion control which was still as WOW as ever. MOP2 was technically more valid in the mobile phone arena, but not visually so dashing, so it was omitted from this first rendezvous. The usual routine: J-P showcased our original demo device to Harri and gave a spirited sales talk about the necessity of novel user interface methods for the 'new century', after which I commented briefly on the technical aspects of the demo, demonstrating viably my superhuman technical knowledge about motion control. As usual, I let others do the bulk of the talking, I still was very much the silent type in this kind of 'meet and greet' meetings. Harri was clearly an intelligent bloke and immediately seemed to understand the possibilities of such user interfaces. He was also extremely talkative and had connections all over the place, mainly from his Sonera-time. In a short pow-wow after interview, J-P, Jyrki and I decided Harri would be suitable to manage MyOrigo to its bright future. Every ship needs a figurehead! So Harri was hired, and would start working as MyOrigo CEO in the early spring of 2001. I would relinquish my original multi-role as the company's acting CEO and become a full-time CTO, Chief Technical Officer, which was of course where I belonged. I was no social maestro, no suitable figurehead for a company that needed just the right presentation to hit the jackpot on the big deals.

Back in the lab, I was moving on fast to the next prototype version – the last one, before the mass product, if all went to plan. MyOrigo Prototype 3 – MOP3 in short – would be a device very close to the final phone that was to be mass produced. Well, its shape and size would be different, but the hardware would be as close as we could get. We decided to build MOP3 around 4:3 PDA display again for the quite practical reason – more narrow, say 16:9, displays were simply not available. It would mean a PDA-shaped 'wide' rectangular device, not really a phone-looking thing at all.

MOP3's specification was simple: the MOP2 + GSM phone hardware and functionality, all built on a single new circuit board designed by Juhani, packaged in better mechanics. We planned to make a dozen MOP3s, so my software developers could each get their own even after management had grabbed theirs for demo and show purposes.

Another three months of design work, building the prototypes, and Juhani had once again outperformed himself. MOP3 was a work of art to my eyes – a matte black, curved casing with alien-like hardware shining through the transparent plastic of HaptiTouch. It worked, and it contained the hardware architecture solutions that would be adopted for the final mobile phone design directly, almost as-was.

MOP3 from the spring of 2001 shown with its HaptiTouch mechanism clearly visible.

For me, the MOP3 confirmed that all the key elements could be packaged into a single phone-sized device. For others, it was really the first physical proof that a phone like this could be made, and actual phone calls be performed with it...

Virtual Mirror got its new demo version on MOP3, smoother and more convincing than before, but it was still all very demo-like. No real software architecture was designed yet, even the final software platform (operating system) was to be selected. But all that was about to change in the coming months: MOP3 provided the hardware close enough to the final mass product that a real, systematic software development could be started. There were enough proto units, around 10, so that each software developer would have one to run their code on.

I believed that the prototyping was now done – it was time to make the final jump to a mass production design, and richen' up to a global gangsta mogul level with the proceeds.

9.
The original 'iPhone', MyOrigo myDevice is seeded (2001)

The best six months of my life had just passed. I felt naively omnipotent after prototyping and tech testing had gone nearly perfectly. My self-confidence was now, finally, at a level that would never be broken no matter what the future would bring. Text-book narcissism and my general incompatibility with any 'company rules' were minor issues I was sure I could overcome and tone down. I was ready to deliver.

I now knew it could be done. A complex machine, a world-changing fully internet-capable phone combining some existing and many novel technologies would be born. And despite all the underlying complexity, on the surface it would look simple and normal to the people buying it. A minimal, yet all-powerful user interface that would become familiar within minutes, without even consulting the user guide.

The long period of button-dominated phones and their usability issues would soon be forgotten. Welcome to the true internet, the full front face touch screen, the freedom to be connected 24/7!

This kind of 'free rein' project is extremely rare in the industry. Usually even before the design team is selected, a pretty detailed specification of the 'phone to be' exists, laid down by potential customers and company management. In practice this means that many features of the user interface, 'old favourites', have to be kept... thereby typically limiting the innovative aspects of any new phone to slight tweaks in the device, new buttons here and there, or a better camera. These projects would never spark a revolution!

But we were not there to maintain the status quo. The parent company Microcell was doing that, designing phones to be just a little bit better, a little bit smaller and faster than those of the OEMs like Ericsson. We were the inventive arm, the not-by-the-book company, the stalking horse of the mobile industry. Appointed by the Godfather Jyrki Hallikainen himself, we would do everything I felt was needed for the revolution, with no regard paid to the old tech facts and idioms. The once-impossible dream of 'creating' my own inventions, to be sold to consumers, was on the verge of becoming a reality.

It was the ultimate delight for an inventor: a moment of calm before the storm, a pure space of invention ideology before implementation starts with all its grinding compromises and never-ending stream of details.

The basics were clear and had been for months in my head. Now I could present them with the technical detail needed to get the buy in of the engineers. Fellow managers and the Board needed no persuading, Jyrki had cleared the decks for anything I had in mind, and Harri was quickly growing to be my kind of CEO, leaving all the technical design authority to me.

My to-do list was simple. Design and build a phone-like device, to be used with mainly one-handed, comprising from the user's viewpoint:

- A full front face touch screen with QWERTY keyboard, all finger operable, no stylus

- No physical buttons except on the sides

- 16:9 wide screen display to keep the form factor 'phone-like'

- A virtual screen to browse, pan and zoom full web pages in real time

- myBook™ page-to-page swipe menus and applications

- Auto-orientation™ at all times, based on position, portrait or landscape

- The capability to install new applications securely

- Better than average capability to make calls, write messages and take pictures

And so the modern smart phone ideology was defined. The actual detailed technical definition, 'the spec' of the device, would be such that many seasoned veteran mobile designers would roll their eyes upon seeing it. That was to be expected from the know-it-all's – "too risky, never been done before, exotic technology..."

I knew already quite clearly what kind of processing power and sensors were needed. In conjunction with the ground-breaking full-web-page-on-mobile UI ideology to support the best possible internet experience, the data connection was to be the best one currently supported. No suitable display existed, so one had to be custom ordered.

The only processor available at the beginning of 2001 which was powerful enough to pan & zoom the virtual screen (1024x1024 pixels) web page in real time **and** handle the actual web page rendering to said screen was Intel's 32-bit StrongArm, an early Arm-based mobile processor. The features that made StrongArm so 'strong' were its large internal caches. A 200 MHZ clock and a wide 32-bit memory bus, all previously unheard of in mobile processors. Intel, after seeing our early demos that utilized Intel's SA-1110 development kit using the same StrongArm processor, were very supportive. They provided top secret documentation and early bug data of the processor, to help us build an efficient design around it.

Memory was the tricky part, and to my mind, we made it too small. Back then, the typical hardware engineer's view was that 'whatever amount of memory we give to the software guys they'll fill it anyway, so let's keep the cost down and put in the smallest chip we can get away with'. With about a million and half other important factors to decide on and manufacturing costs of the device rising to unprecedented heights, I made an error and let the HW engineers have their way. This decision, further influenced by the cost of memory chips, later severely hindered the performance and capability of the **myDevice** applications. We settled on 32 MB of flash storage (for programs and some user data) and 64 MB of RAM. The runnable code was actually copied to RAM in runtime to speed things up, so around 40 MB of RAM was left to be used by everything else – including the internet, phone functions, camera and so on. To put this into context, today's web browsers easily consume 1,000 MB of RAM alone just rendering standard web pages.

The display was trickier still. There were some suitable screens available, but all of them had the PDA aspect ratio of 4:3 or even 5:4. We needed a wide-screen display to fill the front panel and keep the device looking like a phone, not a square PDA. In addition, the readability in direct sunlight of those displays was poor, and we wanted to make our phone useable outside too. So, after long negotiations with Philips Displays, our sourcing department warriors Juha Rytky, Markku Virta and Jyrki Portin were finally able to convince them to make us a custom display. It was to be an ultra-modern, trans-reflective LCD with LED illumination. The aspect ratio of it was unheard of at 1.81 (wider than a 16:9 widescreen), delivering an impressive resolution of 320 x 176 (top notch back then...).

The HaptiTouch mechanism meant that four force sensors had to be planted under the transparent plastic covering the whole front face of the device. We already knew quite a lot about implementation, after three different prototype designs - the stand-alone HaptiTouch proto, the MOP2 and MOP3. But how to make it last in everyday use, how to make it mass-producible and cheap enough...we had no definitive answers at that time.

The phone and data communication part (the 'radio' part) of the device was quite easy to decide on, with my background in desktop computing. We could utilize Microcell's earlier design based on Analog Devices' chipset as a simple modem. The speediest version delivering GPRS class 10 communications was selected to enhance the internet download speed to the best practical level mobile networks were providing. The modem would be communicating with the StrongArm application processor via a fast serial link. Just like plugging a Wi-Fi stick into a computer's USB port, we thus had a two-part system with a separate radio communication module 'modem' and an application CPU module 'computer', one of the first designs of its kind in the mobile industry. The benefits included more power to keep the application running, since the CPU would not have to handle any radio communications, and more standby time as a phone, since the CPU could sleep and the modem alone could keep up the communications. Of course the cost would be much higher than with traditional one-CPU solutions, but hey, we were making a totally new thing, it was not supposed to be cheap!

Finally, the motion sensor part. The only practical choice back then was a 2-axis accelerometer with a relatively low measuring range, so the resolution would be sufficient to support the Virtual Mirror. 3-axis sensors were not available in a single chip, so cost and design considerations dictated that one axis be left out. I intuitively knew then that the mirror would work with 2 axes, so it was no great sacrifice.

After a few weeks, the full spec was done, with the earlier MOP2 prototype we built with Juhani serving as a backbone and proof-of-operation. The engineers were for the most part bought in to it, although some complained about the 'totally immature and unproven' approach to building a phone.

I ignored the naysayers, fired a couple of troublemakers and went full steam ahead.

And so began the design of my unique idea of a *different* phone, the **myDevice**!

10.
Software!

Early months of 2001 turned into a bright spring. MOP3 was working, myDevice was defined by me, hardware was selected and the first ideas about the shape of the 'thing' were on our tables as wood models, 'dummies'.

What actually defines how any display-driven device feels? The software. Modern people start to be very aware of it, comparing iPhone and Android devices they understand that main differences are done by the software. They feel different because of different software architectures, different user interface idioms implemented by software.

It was finally time to plunge into the part of myDevice I was the most expert in. What would be the software platform for MyOrigo's product, which operating system, where to source the applications like web browser?

I initially studied Windows CE – the mobile effort of Microsoft at the time – and decided it was no go. First of all, it was designed to be used by a pointer-like device or stylus. Secondly, replacing all user interface elements to our own versions of them, changing also the whole UI ideology to support myBook, virtual screen and such, would be practically impossible. Third, it was not meant to be used in a phone – a PDA maybe, but features required by a voice phone were lacking. It was like buying a used house that you have to completely demolish and rebuild to suit your taste - it is easier, faster and cheaper to build a totally new house. I felt the situation was same with Windows CE. Better to start from scratch.

All the other commercially available 'mobile phone operating systems' of the time were designed for the 8- or 16-bit traditional mobile processors and low level functionalities, so they would not be able to support sophisticated, sandboxed software architecture I had in mind.

I had a long history of Java coding – most of the work I did on my previous, only 'normal' job I ever had, was done in Java. I was one of the few user interface experts that knew how to implement a smooth and fast UI code with Java. I liked that young programming language and preferred it greatly over C++, let alone C. I had wondered from the beginning of MyOrigo, whether it would be possible to make a fully Java-based phone. Sure there were possibilities to run Java MIDlets (sort of mini-applications, small games etc.) on some phones of that time, but that was only an extra – all the built-in applications and functions were implemented with traditional programming languages like C. Using C resulted easily in horrible spaghetti code – each software module was dependent on many others, and if one changed, others had to change too. Building an end-user ready product with tens of software developers' C code unified into one device was like building a house of cards. Any twitch in any one software module and it would all fall apart, fail to function. I had seen that all too well in Microcell's traditional phone development program – the resulting code was so fragile that it was a challenge to get it working at all, albeit that the functionalities it provided were quite simple. For a more complex device it would be impossible, I thought.

After some searches with AltaVista (Googling was not yet the trend back then...), I found two companies that maybe could provide us Java platforms powerful enough to run a full web browser in Intel's StrongArm processor. They were a Swiss company called Esmertec, and a UK based-company, Tao Group.

After the numerous technical aspects were compared and studied, I selected Tao's Intent platform, that would act as a native operating system as well as an extremely high-performance Java compiler. That was to be the MyOrigo's device software foundation.

In the midst of this, two of the company's futures most important software guys were recruited to MyOrigo. Our CEO Harri's old friend, David Narraway was the first one. He was a smooth British fellow, maybe ten years older than me, married to a Finnish wife. From the first discussions I had with him it was obvious he was radically intelligent man, quick as a lightning to pick up on ideas and work on them. He didn't have any experience in Java, but it didn't matter – that language is easy to learn coming from C++. He was hired with double the salary compared to any other engineer in the company, but I and Harri thought it was worth it. David would be the leader of the software development, at hands-on level.

David had a friend at Nokia, a software veteran and quite a stoic fellow called Jari Turunen. In addition to being surprisingly open to new ideas in mobile SW development (he had seen at Nokia how the old style failed in many aspects, even more so when more complex 'smart phones' came to play), he had a sort of a feet-on-the ground sanity that I liked. I was sure he'd have the capability to understand and force my novel software architecture ideas for the rest of the SW engineers, I thought. So we hired him too, to be the software architecture manager, also acting as application SW manager.

Platform selected, Java chosen as the preferred language for the applications, the key software maestros in the house. Get ready... set... go!

Mobile phone software consists of layers. The lowest layer, just above the hardware, is called HAL - Hardware Abstraction Layer. That layer makes the selected platform software and compiler operate with just that configuration of hardware, memory chips, processors, and so on. Usually HAL is pretty small piece of code that just serves the operating system kernel hide the differences between different processor and memory configurations the operating system runs in - in our case, it would be the final link between Tao's Intent operating system and Intel SA-1110 processor.

Next layer up - the system software layer. The operating system (in our case, Tao's Intent) is running here, as well as the device drivers: small pieces of high-performance software that actually control the hardware components and provide a generalized interface to the top layers. A typical driver software piece could control a display chip with its specific timings, sequences and buses, transmitting the upper software layer's defined display content to the actual display chip in the form that that chip understood. Now if for example some future next-generation device would have a different display chip, it would be theoretically enough just to change that driver software module, all the other SW could remain the same and work just as it worked with the old display chip. Well, you know the drill: when you plug in some new component to your PC, a new printer for example, a software driver must be installed too to manage the actual connection to that printer and interpret the general printable data to the form of specialized data that just that printer type requires. It's the same thing with device driver software in a mobile - all major hardware components require their own drivers.

From there up, comes the final layer - Application Software Layer. It contains the general application framework libraries and systems, user interface software libraries, and the actual application programs (user level programs).

In addition to those, we had Intent's Java compiler sitting in between system software layer and application layer. Since all the application software would be written in Java, Intent had to compile it to the processor-native machine code that could be executed on and interfaced to the System Software level OS and drivers.

In addition, there is always a small piece of code called the Bootloader. It is the code first executed when a device starts up, it usually checks if there is a flasher connected (a gadget used to reprogram the device), and if not, it calls the boot section of the operating system followed by a general start-up call to system software, starting the actual booting procedure of the device.

Our system software layer would contain one interesting addition to the typical driver structure: a virtual screen driver. That would be driver software to emulate a bigger physical display than we had, and then showing part of the screen zoomed and panned around the actual physical screen. Fooling applications into thinking they actually had a pc-sized screen to draw on... Other than that, SYSSW layer was to be pretty standard stuff. My original and faithful System Software team, first software resources mainly consisted of my old friends, would take care of that. Now when the platform was selected, they could start to study it and make it work on the MOP3 prototypes, which were close to the same HW configuration the final product would be.

Application software layer. This was the thing where most of the new magic would happen. I wanted to revolutionize not only the way that mobile phones were used, but also the way how

application software was written to them. I
felt the style I had witnessed at Microcell –
gluing together many highly case-specific
pieces of C code and trying to make them compile
as a whole – was like torture from the middle
ages. I had done PC programming for 15 years,
Windows programming with Microsoft's Visual
Studio C++ and J# for a long time. I knew there
were better ways to go about it in mobiles,
ways used in sophisticated PC software for
years.

In our long sit-downs with Jari and David, I
laid down my idea of our future application
software architecture:

There would an Application Framework, a sort of
combined post office and police of the device,
that would handle all the transmission of data
between the applications. The framework's key
mechanism would be InfoBus that would broadcast
Events provided by applications, in effect
removing the need for a direct communication
between two applications in the device. The
framework would also provide applications the
possibility to start other apps, so the
myBook(tm) book-like menu structure could be an
application itself too.

InfoBus would generalise and make robust many
tricky operations that required applications
communicating with each other. A classic case
is that the phone rings before the complete
address book from SIM card (yes, back then phone
numbers were actually stored in the SIM) has
been read into the device memory. Now this means
that when the ringing starts, the phone knows
only the number of the caller, not the name.
But couple of seconds later, when the address
book has been read from the SIM, the name tied
to that ringing number would be known. But...
the address book is one application, phone dial
application is other. In old style programming,
the phone dial app has to have specific
knowledge of which function to call in the
address book app to get the name related to a
phone number. And because the address book app
does not know it yet, there has to be mechanism
for the address book app to tell the phone dial
app that „I don't know yet, I will soon, maybe".
And then, the phone dial app must be smart
enough to call that function again to find out

the name of that caller – after the address book app has finished reading the SIM card. And Lord help us if anybody changed the code on the address book app or phone dial app after this complex hassle of interaction function calls between apps has been made to work. Any change can result in the effect that neither the phone dial app nor the address book app works correctly afterwards. In the Windows and PC world, they called a similar effect of applications and software libraries depending on each other's versions "dependency hell". In the mobiles of the 2000's, it was much worse than that name suggests!

With InfoBus, the flow of actions would be quite different in that same case: the phone dial app sends a general event to InfoBus, stating that a phone call arrived, bearing certain data like the phone number, start timestamp and so on within the event. Now in that general event is a field for the caller's name – which the phone dial app leaves empty, it does not know it, it doesn't keep database of caller names. The address book app receives that event from InfoBus, recognizes the general format of the event, and sees that it contains an unfilled field for the caller name. The address book stores the event and waits until the SIM card read is finished – and then simply matches the name from the address list to the number on that event that was stored. Now the address book app does not know anything about the phone dial app – it does not have to know – it only filled info into a general event that arrived from InfoBus where all the applications are connected to. The address book's job is done; it goes about its own business. The phone dial app sees that the event it sent (which it still has a reference to) has been changed. It re-renders the info that event contains, and voilà! – the name of the caller appears on the phone dial app! The phone dial app never had to know anything about who actually provided that caller name, it only sent out the event and got the info it wanted ubiquitously.

This way InfoBus made the applications and their versions independent of each other, as long as the base event structure was kept the same. And it is much easier to keep event structure stabile, especially because new events could inherit from the old event types we would define in the beginning of the development. So any application could be updated freely, it would not affect the function of other apps – if implemented properly (that's the ultimate trick of course, but all in all, this system worked quite well.).

Most apps created by us would run in a single memory space, in other words, unprotected from each other. They would have sort of "superuser" capabilities on the device. It was also practical for them to run on a single Java Virtual Machine – so they could directly share the Events over InfoBus without the need to serialize and de-serialize those Events for each application. The downside was, badly programmed apps could collapse the whole device when running in that single VM. So we would have to be careful about which apps we allowed to run in there.

External applications, downloaded OTA (over the air) or from a memory card, would run in a sandbox. Because applications were implemented in Java, this was quite easy to do – Intent supported creating an own virtual machine for each Java application, and in effect, that was the sandbox mechanism. Limiting the capabilities of that specific VM running external application would limit also the application, no going around it. So, even if malicious, downloaded apps could not corrupt the device – all they could do would be the bang the walls of their limited sandboxes. This is the key to providing the possibility to download 3rd party apps to a mobile phone – keeping the device safe. MyDevice was to be the first phone in the world that enabled real, downloadable full-spec apps to be run. At that time the state of the art was mini-applications, Java MIDlets, that could at most implement a simple game on a mobile, not much more. All that would change with us...

After all the technical talk was done, the structure of the software was decided quick with those master coders. Especially Jari seemed to be excited about this sophisticated architecture that was never done to a mobile phone before this. Java was also POP for him – it was the language of the future, and now he got to be the first guy in the world to implement a fully featured, ultramodern Java-based application framework to a mobile phone. A breath of fresh air after Nokia's cramped and complex C++-based old-school Symbian OS platform!

11.
The First Negative Chapter

Prototyping MyOrigo's visions, building and testing MOP2 and MOP3, and finally locking the specification of future mass product were on the winning side of the flip coin of my life. Also the software work had started with real enthusiasm and the s/w resources were absolutely top notch, world -class.

But there was also the dark side.

Beginning already in 2000 and progressing even deeper after Harri became the CEO at the spring of 2001, I slipped further and further away from society's accepted norms. I now slept to the noon almost every day, some days not showing up at all to the 'crime scene' (MyOrigo office), as Tapsa called our headquarters. I felt it was my prerogative, I could act like this when others couldn't. It seemed to work – on the surface. But arrogant and ignorant behaviour towards others, showing a bad example about letting work hours' slip and skip, had an obvious effect in my colleagues. In the spring of 2001, there were already almost 20 guys working for MyOrigo. Many of them developed bad habits, obviously copying mine – some had short working days arriving at the office around 10 am, leaving just after 4 pm.

Of course nobody ever imitated my better workmanship

features in their behaviour (pushing work through the night when some crisis required it, or inventing new technical solutions that actually worked on the fly), just the conveniently bad elements they felt they were entitled to – after all, they also worked for MyOrigo, and if Johannes behaved that way, why couldn't they? So they did.

Laziness was only one of the real problems, and certainly not the biggest. A real resistance towards my technical leadership started to form at MyOrigo. The guys who were hired from various positions of the existing mobile industry – from Nokia and other major companies – did not adapt to the agile working methods of our small company. They just wanted to do their job, and nothing more, then go home and watch football on the telly. If any job was to be done unconventionally, they would surely feel insulted – they knew best how things are done in the mobile industry! Unfortunately, the totally new architecture of the myDevice was not be implemented by methods they knew, their experiences from designing 8-bit phones with b & w screens made them outdated for this kind of project. In practice we were building the most powerful hand-held computer possible at that time, and it required a new mind-set for every designer.

I knew if we made the myDevice like these guys would have it - old school - there would be practically no new technologies, no new innovations, just a crooked copy of a Nokia Communicator (the so-called 'smart phone' of the time). That would render all my bright ideas in vain, so I pushed hard with my requirements and implementation architectures. The know-it-all's - senior engineers - pushed back as much as they could.

The newly hired project manager that handled the everyday bread-and-butter design work of myDevice,

Aimo Vainio, was an OK guy by any standard but certainly too soft when listening complaints from the engineers. This was partly because of Aimo's Nokia background – he knew only too well how things were done at Nokia (and hence should be done in any other company too), and engineers were able to justify their slowness or unwillingness to comply with my work definitions simply by appealing to that side – the Nokia side - of Aimo's mind. "It has never been done like this and it will not succeed...".

The first real manifestation of this resistance was the exhaustion of Juhani Putkinen. Juhani had done a ground-breaking job so far, but our new hardware engineers did not like his working style. Juhani was commanding how to proceed in military style – others had a certain timeframe to express their differing opinions, and if they could not argue their viewpoints, things were going to be done Juhani's way.

After half a year when Juhani made a clear design error (after a hundred brilliant and successful design choices), the cry-baby HW engineers had the winning ticket in their hand. They complained to Aimo about Juhani's working style, and naturally Aimo went over my head to bring those complaints to Jyrki, the godfather himself.

Juhani was relieved of all design responsibilities at MyOrigo, work was to be continued without him. I wasn't really even asked if that was right. It was my first real loss, both authority-wise and resource-wise. I now knew I had to stand my ground on authority, build a social network also inside the company that would support me, otherwise I would soon be exhausted from my own company! But how could I do this when my working style was so irregular – most of the time I wasn't even at the office.

By extreme and unlikely luck, both Jyrki and J-P understood my value then, and up until the end. They felt that the implementation of MyOrigo's brightest new ideas would fail without me, and so at any critical moment, I always got the final support of those fine gentlemen. But there were glitches in the support. That allowed the resistance to do their handiwork – Juhani was the first casualty of that war.

Later in 2003 when MyOrigo's myDevice mass product was practically ready and in trial production, there was a serious attempt to rule me out of the equation, supported by many senior employees of MyOrigo. The company's third CEO, Matti Paasovaara, had decided to drop me from the board of managers after listening to several complaints about my "tyrannical and politically incorrect" working methods. That would mean in practice I would be isolated to some dark corner of MyOrigo to 'develop' my ideas alone, not hindering the work of 'real engineers'. In a way I thought Matti's plan was very understandable – he was not a technical expert so he couldn't understand my critical role in the development of the key new-tech pieces - and he had mixed feelings towards me from the very beginning. I had treated him with a touch of hostility from the day we first met, in my youthful arrogance, now it was the payback time. Luckily I already had heard about Matti's plan, so I was prepared. As I stepped into Matti's office and he told me he wanted to remove me from the board of managers, to demote me to reporting into Aimo, I handed him my signed resignation. Matti did not expect that, he thought the company, MyOrigo, would be my life that I could not give up, and would be ready to compromise anything to keep working there. He was wrong – he did not understand what makes an inventor tick. When he understood the contents of my resignation, a sudden flash of disbelief twisted his face for a second. Then started the explanation: "I am only considering this possibility, I have not yet decided that I will actually downgrade you", and so forth. I walked out of the office telling him he had till the end of the day to state his intentions, then my resignation would become valid.

What happened after behind the scenes, was only told to me years later. That same day, Matti had requested a urgent meeting with Jyrki and J-P, stating his viewpoint that I was hindering the company's future and all development efforts with my arrogant attitude, and I did not take into account any suggestions the seasoned veterans gave about building a mobile phone. I should be excluded from the company if I did not comply with Matti's idea of putting me where "I could do no harm". That meeting had ended in Jyrki stating bluntly: "Matti, now you have to decide if you want to stay working for MyOrigo or not. Johannes will stay on the management board for the foreseeable future, and he will lead the technical efforts as before. If you can support him in that you can stay."

Later that day Matti informed me that his idea wasn't very good, let's keep working the way we have been doing. I did not think about it any further. I continued my work and put that whole episode behind me – I was quite gifted at forgetting the inconvenient incidents of the past.

When I began working for my own new company, I wanted to hand-code a lot of things for myself, but after a year I had become a lazy coder. I rather just advised others and 'managed' the critical development pieces, like HaptiTouch (that was constantly giving headaches to mechanics team) and the myBook menu system. The only thing I actually coded myself was the motion control software that did the math for Virtual Mirror, algorithms and filtering of the motion sensor signal. It was a neat piece of code, but I could have done much more. If I should have, is another question entirely – everything got done after all, about a million lines of software code in total.

And yes, as I am human, there were some people that just generally pissed me off. One guy called "Puuha-Pete" (Pete the handyman) was copying presentations from others and later showing them as his own ideas. In addition, to any new idea I brought up, he always combed the internet to find if anything even vaguely similar had been thought of before to be a killjoy. On top of all this he never actually did the tasks he was supposed to do, working under me or others. The senior guys objecting to my working methods I could stomach and had no real emotion towards them, but Puuha-Pete I did dislike on an emotional level. It was rare for me – as a socially limited person, I usually did not have strong emotions towards people, positive or negative. I did try to fire him, but he was expert in legislation regarding that, unsurprisingly. Finnish laws back then were very rigid and practically prohibited firing people from the employee's committee, and Pete had wisely placed himself on that committee from the beginning... I am not sure what Puuha-Pete got out of working in our company, he had a CV which meant he could get work in some other company easily, it was the mobile industry hype years in Finland.

One unlucky guy had it even worse, he got my everlasting hate for a small thing he did at the new employees' presentation meeting. In our hottest hype we hired many new people every month, so once a month there was a get-together where all the key managers presented themselves. This time that new guy, Sami, was there, hired without my help in the process, probably recommended by an earlier recruit. As I introduced myself in my unsocial jerky style, I finally stated that "as you all probably know the design work here is based on my inventions". This guy laughed aloud! Shortly, but snappily. Nobody else did and he stopped laughing very quickly. I never understood – and still don't – what motivated him to do that. I have hated him ever since, even pushed him out of a later design project year after that. The poor guy did not really deserve that – I don't know what tricked my long-lasting negative emotion towards him, and I never cared to analyse it.

I was also getting fat. When we established MyOrigo in September 2000, I weighted 90 kilos, a nominal figure for my 183 cm height and heavy build. By 2003, it was up to 105. I had developed a taste for fine dining lunches and dinners, accompanied by French and Spanish wines. Drunk at least 4 nights a week, my body started to protest not only by gaining weight, but also in the form of constant throat burns. "The side-effects of being a modern day inventor", I thought and kept drinking, also all the more often heavier stuff like Calvados. I was not in very good shape physically, but my body was still able to take it – I was young. But youth, in the mobile industry, doesn't last very long...

12.
myBook:
the original "swipe"

In the midst of defining our to-be-revolutionary Java software architecture during first half of 2001, I was also digging into the tedious details of the book-like main menu system. We had decided with J-P that myDevice would not contain a "traditional" tree-like menu system at all. The book concept was there, sure, had been for months then, but how to actually implement it, what to really show on those virtual pages... there were a thousand technical nuances to speculate and decide on. Make a book-like user interface, that is also active for many functions other than reading? Easier said than done. First I needed to understand in more depth what is actually so comfortable in a real-world book, then try to capture the suitable elements into an implementable spec.

The most natural form of browsing written or pictorial information for us humans is a book. Not e-book, or a book-like presentation of a single page, but the whole concept and user interface of a physical, traditional book.

It's not only that the individual pages are just the right size for human cognitive capabilities to intake with a single gaze. It's the way information is organized. There is a certain beginning – page 1 – and end. The pages are

---

always in the same order, so the same page will be found in the same place every time. Information is not in a tree-structure but on a single level, in that one stack of pages. That makes it easier to remember where something specific that is looked for lays – in the first quarter of the book, or in about the middle of the book...

Browsing a book is so far the best method designed by man to go through a large quantity of information and find the specific part that might interest you.

All the information and functions the user needed would be organized into a stack of pages – a book. Book pages could contain links or buttons that would lead to other pages of the book, that was a given, but otherwise the idea was that user would feel like they were browsing a normal paperback book. Only this book would not contain a novel by a young, suicidal Russian author – it would contain the user's short messages, their contacts and calendar, and their ring tone settings... all in those same stack of pages, all in a never-changing order.

I really liked the idea of this kind of digital book in place of a menu system. I thought the concept was big enough to give it its own name. Everything in our company seemed to have a my-presyllable, so myBook would do. Oh my.

As soon as our old-school mobile software designers had understood the virtues of myBook, they all liked it – nobody wished to go back to the "Nokia tree-like menu style".

myBook could have never been implemented without HaptiTouch finger-useable touch technology. the critical element of the book UI was to be able to turn the page with a natural motion – flick the visible page from left to right and page would turn, showing the next page. Just like in a real book. This was the first presentation of the famous "swipe"-motion Apple so keenly wishes it had implemented first. But no – swipe motion was first used in myDevice menu system publicly as early as 2002, in prototypes already during 2001.

myBook's first couple of pages, replacing the traditional main menu, were logical and easy to understand – like an index of a thick book, listing out the things that the underlying book contained. Things like Calls, Messages, Internet, Settings, and so on... Swiping between those first couple of pages felt just like flicking the pages of a real book.

myBook presented not only advantages, but also challenges to the application design. For example, how could Virtual Mirror be supported in the pages of a book? We decided it couldn't, so certain applications like the web browser and documents viewer popped out of the book pages and took the control of the whole screen. This was also a lesson to the stubborn one-idea pushers – some cases just require exceptions to be made into rules, even with the best inventions. A borderline case was the Camera application, which we finally decided to implement as a page of the book.

In almost all standard applications, like Calendar, E-mail, Message viewer et cetera, myBook was more than perfect. Applications only had to decide what to show on each page of the virtual book, the application framework with integrated myBook system took care of the rest. The message application did not have to figure out how to layout or scroll through 50 messages, it just added 50 message-filled pages to myBook, and all browsing and viewing user interface functions were provided by, with no extra code needed in Message application. Application design and implementation was a pure joy compared to any old school system – hell, it was better than current (in year 2014) Android or iOS development environments!

The actual implementation of myBook, first to MOP3 protos, then to MOM1 (myDevice) pre-production units was not without its challenges, as always I wanted high frame rate to make the turning of the page look fluid and smooth. The smoothness of myBook got better in many iterations during 2001-2003, but never reached the 30 FPS (frames per second, a measurement of how many times per second the image on the screen is updated, translating to how smooth the display action looks to the user) stabile level that Virtual Mirror-based applications like web browser were able to run on. I was disappointed but never got around in actually trying to improve the performance of myBook underlying rendering code – I thought it was good enough for now, because it seemed I was the only one complaining about not-so-smooth page turns on the screen. It seems that nagging and grinding compromises creep in whenever something this complex is made.

The detection of a finger swipe from right to left (and vice versa) was surprisingly difficult, too. MOM1's final production version's HaptiTouch performed beautifully with a finger pressed on a static location, but the finger swipe created a lot of friction on the surface of the HaptiTouch plastic panel, and distorted the touch location readings severely. It was David the Master Programmer who finally came up with a tolerant and quite robust algorithm for detecting the finger swipe.

The book configuration of most of the build-in apps opened up a unique opportunity for user interface flexibility: MOM1's displays aspect ratio was close enough to 2:1 widescreen that each page could be viewed horizontally as well as vertically. The items in the page would be divided to two equally sized vertical blocks when device was used vertically, whilst the horizontal view showed those same blocks on the left-and-right side of the screen. Some scaling was needed but little enough that UI design could be the exactly same for vertical (portrait) and horizontal (landscape) orientations of the device.

Since we had the motion sensor data available anyway, it was simply a logical addition to make the screen automatically adjust to portrait or landscape modes when device was turned horizontal or vertical by the user. This feature requires no further explanation today in 2014 as every smartphone in the world now implements this same feature in many of their applications – the screen turns automatically landscape/portrait when user turns the device accordingly.

When we did it there was nothing like it in the world, in any other device.

We decided with David to call that feature "auto-orientation" of the screen.

I still miss using the myBook system in those myDevices. Every once in a while I dig one of the old production units out of the box and flick a few pages. That's about as nostalgic as I get, ever.

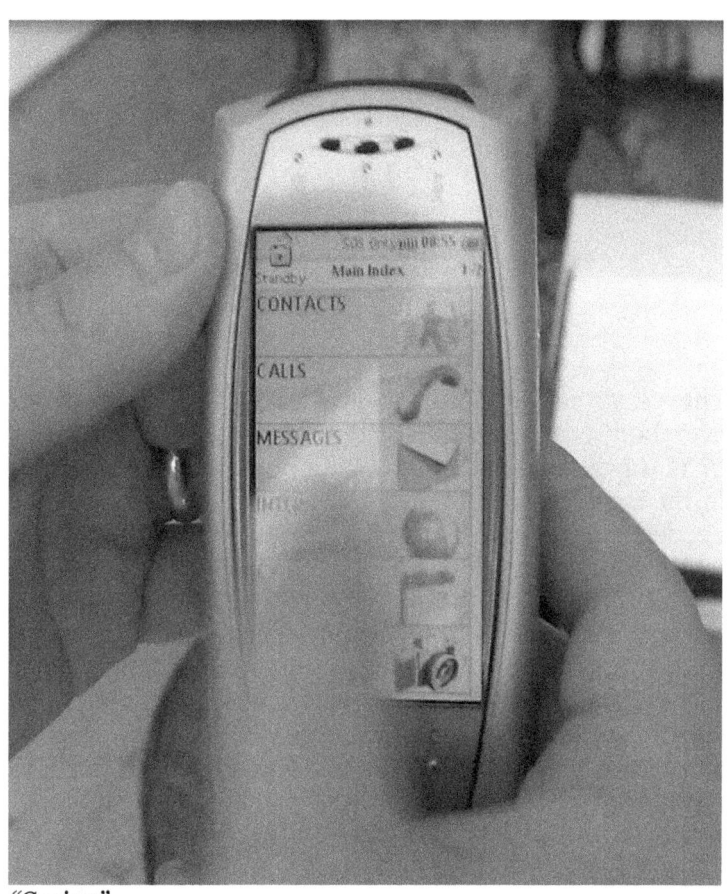

"Swipe"

13.
USA

They say that in America, everything is bigger – and everybody can become a millionaire. They say that inventors are valued in America more than anywhere else in the world. It was the time to put that to the test – we needed more money for MyOrigo, and what better place to get it than from the USA.

I was the CTO – chief technical officer - of MyOrigo, in charge of all the avant-garde technical stuff and talk. There were situations where some external party had to be convinced that we knew what we were talking about technics-wise. Harri was a great salesman and a slick-mouth-johnny for sure when it came to company presentations and sales pitches, but he could not go deep enough in the technical details with some of the potential partners. So I had to participate, as if the software architecture, myBook structure and tinkering with HaptiTouch mechanics weren't enough to keep me occupied...

Jyrki had secured MyOrigo's funding to some extend already before spring 2001 by circulating subcontracting money via Microcell to MyOrigo, but it was too ad hoc, and would not last, he told. It was time to find some hardware supplier to pay for the development of myDevice, in Microcell style. Microcell had ingenious funding principle in the beginning: Jyrki promised National Semiconductor access to the very lucrative mobile phone component market, if they funded the design of the phone using just those components. And since with his connections he had already a letter of intent signed with Ericsson – a top tier mobile phone brand back then – National Semi was all too eager to believe that Jyrki could deliver what he promised. So Microcell got funded by the actual seller of the electronic components that their designed phone actually needed to get build! Bold move, and it had already worked for a year.

Our most prospective vendor regarding possible funding was a large, traditional American semiconductor company called Intel – we had used their CPU in our prototypes already, and at the same time, Intel wanted to push into the mobile phone component business where they really weren't at the time yet. Intel would want to see their CPU in any mobile phone, let alone as advanced as we were up to.

I had my new passport, new thousand dollar Italian Barracuda-shoes, and a poorly fitting Hugo Boss suit (my body was never very suitable for those ready-made suits, I had a hint of a hunchback from sitting in front of computer screen for almost twenty years) when I took a taxi to the airport. Juhani Putkinen, Jussi, was part of my gang, my trusted HW designer, and since we were going to talk to the HW vendors (the professional sounding jargon-name for electronic component manufacturers in mobile business), I thought I needed him to convince those Intel guys we were the real deal. Harri and Jyrki were already in the States, hustling around on Microcell business – they would join us for the most important of MyOrigo meetings.

Even years later, I never learned to like those long-haul

flights, business class or not. There is always that faint smell of disaster in the air, and I don't mean an airplane crash; I mean delays, missing luggage, overbookings, jetlag, insomnia, border control and customs procedures (which can be 'interesting' when you are carrying prototypes with wires hanging from them), and so on. This was my first trans-continental trip, and I didn't know how to prepare – I even had checked baggage in the cargo against any seasoned traveller's advice. Since we were starting our journey from a small city, Oulu, the typical route to Austin (our first meeting site) would be something like Oulu-Helsinki-Frankfurt-New York-Austin. It's easy to imagine how many things can go wrong with such a multitude of hops between endpoints.

We tried to spend our time on the plane productively with Jussi. It failed – I was just about to learn that I could not concentrate on pretty much anything in the air, not even reading a book really. Maybe the numerous air pressure changes between 1.0 bars (normal ground level air pressure) and 0.75 bars (typical pressure on a plane flying at 10.000 meters) were swelling and compressing my brain, physically pumping the sense out of it – I felt I had the concentration ability of a 2-year old. What else to do then, I resorted to drinking and trying to catch a moment of shut-eye.

Also, on even short hops between close-by cities, my skin reacts radically to flying. Not a rash or anything medical, but it starts to smell exactly like – old kerosene. I can only speculate from where that odour originates from, if there is a whiff of kerosene in the aircraft cabin air or what, but that's what my skins stinks like after a flight. And even after shower in the hotel, my clothes seem to keep that stench... I am not designed to be a travelling man, it seems. At least not by plane.

Mercifully, this time all the interconnecting planes were pretty much on time, and we arrived in Austin after just about 22 hours of continuous travelling. Customs procedures were interesting – I walked through them without any issues, Juhani seemed to get engaged in quite a long interview in there, maybe he had something in the registers because of his military background. But he got through too, finally.

After a short sleep in our designated motel near the Airport, we aimed our rented Jaguar S-type's nose towards Intel's research and design headquarters at the other side of the city. Sleep cheap, but drive a treat, was my motto, hence the Jag. Ah, and big Citroen's were not available for renting in the US, of course. I drove, Juhani rode shotgun and interpreted the map – yes, it was the time before those all-powerful GPS navs really became standard equipment in any rented car (or phone).

The meeting was to be a simple one. Our goal was twofold: Primarily, to get Intel hooked on the idea of their SA-1110 StrongArm processor inside the most revolutionary phone in the world – and secondarily, to get hardware design support and necessary semi-secret documentation of this new processor's errata and hidden features to make its integration into myDevice smooth and easy.

The Americans welcomed us warmly. Harri and Jyrki were also present, after waiting for our arrival in Intel's lobby for a while. After the short introductions – which I hated then, and still do, I can't ever remember anybody's names afterwards – I presented our MOP2 prototype, it was the best we could show at that time. MOP3 would have been available, but it didn't have SW good enough to demo it yet, so its predecessor had to do the job.

The Intel guys looked at the Virtual Mirror with some

amazement. I though the demo was really not very convincing to a layman, yet, they seemed impressed. I also briefly demoed the (largely misbehaving) HaptiTouch and emphasized that usability with finger, not a plasticy-stickety-stylus. Force sensitivity was something totally new, they could see a circle on the screen expand under my finger when I exerted more force on the plastic lens on top of the display of 3 inch-thick MOP2. Lastly, I explained that this whole prototype was built around their SA-1110 development kit, which seemed to make them happy and chatty. "Good", I thought. We spoke some technical jargon, even Jussi got his say, speaking with stubborn longevity about the internal caches the SA-1110 had.

Lunchtime, and we had not even got to the real business part yet, extracting funding from Intel! Maybe the most disappointing thing about the USA for me was the business lunch culture, at least on that first encounter – some dudes dropped bags of chips, mayo-filled sandwiches and soda cans onto the meeting room table, this was supposed to be a "working lunch". I started to grasp why America is the fattest country in the world, even amongst high-class business people...

The rest of the meeting was a back-and-forth volleyball about how Intel could help us financially. I didn't even follow most of that discussion, I was tired, still exhausted from the trip, and the successful demo show had also worn me down – I was not yet used to presenting our gear to "the folks in the business". The disappointing ending, which everybody still tried to make sound a success from both sides, revealed the fact that Intel had only one instrument they could use to fund us: their newly founded investment fund, whose investment processes were still in development, reputedly. In reality it would mean that at least half a year would pass before any funding from those guys would possible – if ever.

We never got investment from Intel, but the meeting resulted in something good anyway: for years to come we had absolutely the best-quality support from Intel, they provided all the secret documentation and help needed to utilize their tech in our phone design, and sold us their processors at the prices usually reserved only for the biggest Fortune 500-listed customers.

The rest of the trip was interesting, but likewise not successful in getting money for the company. Harri and Jyrki went their way, we continued our tour with Jussi to more technical meetings. We met the Analog Devices folks regarding their accelerometers (which we were going to use anyway), and National Semiconductor in Chicago to discuss their sensor developments.

I got a taste of what it means to be in international business. I liked the insider info that wouldn't be publicly available for a year or more, and the fact that I could show something nobody of these big-time players had seen before. But I sincerely hated the travelling and the constant need to sell, sell, sell your ideas – to get big companies to provide components with the right prices, they had to believe in us. Actually, I was starting to cultivate a little bit of a similar attitude towards the USA on this first trip – I liked the general capitalism, the availability of tech and big V8 cars to almost everybody, I disliked the constant need to be social in the how-are-you-how's-the-family-style.

Something good and something bad. I returned to Finland with Juhani, he continued to his hometown Varkaus, I flew from Helsinki to my home at Oulu and slept for almost a whole weekend.

14.
First sale (Radiolinja) and Sonera spy attack

Harri was an active CEO, and highly connected with the mobile operators. After all, he was ex-management of Finland's largest mobile network, Sonera. His departure from Sonera ended up in a gargantuan argument that lasted for years, and resulted in several media-frenzy court cases. Sonera had never paid Harri the compensation promised, so he felt it appropriate to seek novel adventures. The CEO position at MyOrigo was proposed to him just at just the right moment.

It might have been the opportunity to step on Sonera's toes, or it might have been the general sales skills Harri had – Sonera's largest competitor at the time, Radiolinja, was ready to make a deal with MyOrigo after listening to Harri's sales pitch. It was the summer of 2001, and we were still fiddling with MOP3 prototypes, early software tests of myBook and the application framework. These guys were ready to buy – and prepay – a batch of a final product which did not exist yet. I thought to myself that Harri must be the greatest sales guy in the world to pull that off, back then I didn't understand the rules of the knock-out game between big companies like Sonera and Radiolinja.

As we were preparing for the final meeting where the Radiolinja CEO would sign the purchase deal with MyOrigo, Sonera woke up. Obviously they had heard from somewhere that Radiolinja was about to make a big deal with some secretive new company in Oulu – a company that Harri was leading. What resulted will remain in Finnish company crime history for a long time – Harri was put under close surveillance by Sonera, their chief of security actually harassing Harri at airports and any place they could reach him. I also got to sample that harassment. We were driving to Oulu airport at spring 2001, to make a hop to Helsinki to attend some business meeting. Harri had told me already about being under surveillance and manhandled by Sonera's goons – I thought Harri must be overreacting or blowing things out of proportion, after all, this was Finland, the safest country in the world...

Driving our company's flagship Citroen C5 V6 to the airport parking lane, I saw three men running towards our car. There was a bloated shorty at the front, waving a stack of bint papers in his hand, followed by a couple of mammoth sized guys who looked like Chechen security goons. Harri shouted: "Drive on, they are about to do for us!". After a second of confusion I picked up on the situ and accelerated. Sonera's security chief actually jumped in front of the car and I had to do a sudden swerve to avoid hitting him. I thought... waaatta hell was this. Something like this never happened in civilized Finland – or did it? And after thinking about the episode a bit – how could they know we would be arriving at the airport at that time? Illegal mobile phone surveillance, tracking where we were, or even listening to our phone calls? Accessing airlines' databases illegally to follow Harri? Oh shit, what had I gotten into with him?

We switched off our phones, removed even the batteries -

then we drove straight to Sweden to the nearest airport we found from there, in a small town called Lulea. We figured even if Sonera's guys had tabs on a Finnish airline reservation database, they probably wouldn't in Sweden. It was not quite clear to me what the Sonera guys wanted, but Harri convinced me they were up to no good and would resort to physical force if Harri would not, for example, sign the papers they had with them. All this was difficult to believe for me, but seeing the moves of those goons at Oulu's airport just a couple of hours back made me lean towards the notion that Harri's story could be real.

It was time to go to ponder about all of it for the weekend in some place far from Finland and Sonera. We booked the first flights to Stockholm, found a nice hotel in the centre of the old town, and proceeded to drink a nominal-Finnish quantity of alcohol. The evening was quite unmemorable except for our encounter with surely the most beautiful woman in the world. That specimen was a sales girl at a local McDonald's, wore a dirty white t-shirt and cheap jeans over her coconut-brown skin. If ever there was a gazelle in a human form, that was her. Harri – a married man back in the day – immediately proposed to her, I just stood in awe. She didn't accept...

A misty but bright morning revealed itself in Stockholm, cold white light did not treat kindly the face of exhausted, young businessman in a hotel room mirror. The scene looked like I was 50 years old and in bad shape, when in reality I was just about halfway to that golden age. Well, time to get up and go, no matter what the morning sulk felt like. Rendezvous with Harri at the breakfast coffee; we felt we had to continue our trip to find some further enlightenment for how to position ourselves in this slightly comical, but also uncomfortable situation with Sonera. We took a taxi to the airport and continued from there on the first flight to where-ever – which happened to be Prague in the Czech Republic.

What later happened during those two long days and nights in Czech can only be described as "revolting carnage of over-confident foreign businessmen spending a weekend in a town with a collection of local beauties.". Everybody in global sales knows how these things go, going to a less-than-fully-developed country's main city, before tourism has really reached it with its standardizing, numbing claws. Foreign city... always some drinks, and then an enthusiastic hop to explore the city for the local skirt. With years it gets tiresome and all too familiar, but I was young and inexperienced in this kind of travel. For me it was an eye-opener of the best (and general morality concerned, the worst) kind. I decided to like this liberal lifestyle-on-travel, which Harri obviously knew already quite well.

At the final stretch of our all-inclusive weekend trip when we were just about to leave for the airport to fly home, we decided to have a one last Czech meal in the centre of Prague. Harri knew how to look for the best restaurant available: "If you are in the central square of any major city, you have to pick the smallest, the narrowest road leaving away from the square. Then, walk at least 30 meters, and pick the first restaurant that has a tiny, understated street sign in the local language, and no window advertising. Enter that place, and you shall not be disappointed – also you won't see any other tourists.". Following his guidelines and doing the walk just as described, we entered a small cellar pork-themed restaurant with a lot of local people in it. We were hustled to free chairs around a table with other people already sitting at it, and two Czech language menus were slapped under our noses. We both picked the most expensive item on the leaflet (names were incomprehensible, and online internet translation was science fiction back then...) and waited. And waited. After two stout local beers, pork steaks schnitzels in strong mustard with some cabbage on the side, landed on our table.

It was one of the most delicious meals I have ever eaten. It did not taste like pork, it tasted like a... decadent and drunk heaven.

Back home Radiolinja CEO waited with his thousand Euro ballpoint clicked erect. MyOrigo's first major deal was signed with all the party members present – Jyrki, J-P, Harri, Johannes (that's me), and Radiolinja's top drawer management. It was a big thing to us – they paid quarter of a million Euros before anything was delivered, just based on the demos, protos and Harri's slick talk. And even more based on, as I later learned – the perfect opportunity to step on Sonera's toes, buying their ex-VP's stuff right under their noses.

We were sure this would be only the first deal in the series of many. If Finland's second to top mobile network was ready for this, what would happen when the actual myDevice was be finished and presented to the world? I would be disappointed if less than 10 million units were sold...

15.
The taste of money

After the summer of 2001, MyOrigo was making a steep climb on the hiring curve. It was possible because there was a significant income from Radiolinja deal, in the palatable form of prepayment before we had to deliver anything. More importantly, Jyrki had been able to arrange a new and juicy subcontracting deal between MyOrigo and Microcell so well that we were able spend a lot. That had been necessary in the aftermath of our trip to US; we did not convince any HW provider to invest money in us, so "trick number 9" that had worked for Microcell hadn't worked for us - at least not yet.

Namely – and oftentimes only namely - almost all our employees did subcontracting work to Jyrki's main company with hefty compensation, more than double their actual salaries cost. MyOrigo could build and expand its own operation fast; what had started just a short year ago with one employee – me – was now a twenty-person collection of various talent. I didn't mind that we were growing like a bloodthirsty tick in Microcell's back – I realized it was the only way to fund my company so effectively without selling most of it to the niggling investors. This system was also in Jyrki's best interests. As the biggest shareholder of MyOrigo via his holding companies, he didn't want to dilute his cut.

Recruiting had been hit and miss, that was already obvious. But there were enough good guys to carry the whole company ahead – the rotten apples were mainly just lazy or incompetent, not real saboteurs – *yet*.

The organization was forming into a typical multi-headed dragon. The noteworthy element in company's structure was the lack of sales and marketing people – just two or three people could be considered other than engineering or design resources. And yes, it's true – Finnish engineers, even at manager level, can't really sell, much less market. It was not built into Finnish culture, marketing skills. Not like in America, anyway, where everybody must learn to sell themselves even to get a job... No, in Finland it was enough to be a good engineer. Talk was cheap, results were the king. To a typical western world guy that might sound strange, and in today's view, it of course is.

The reasoning behind so few sales guys was primarily money – there was just enough to do the design work way I wanted – but also the company's sales strategy. Our master plan was ODM – Own Design and Manufacturing. We wouldn't own a brand, or market anything to the consumers directly. We would sell to the big brands, where a couple of closed deals meant already millions of sold devices. There would be some mammoth logo – like Ericsson – on the forehead of our phone, and no consumer would ever know who actually designed and manufactured it. There shouldn't ever be any need to do marketing or sales to anybody below tier 2 in the mobile business playground. So a couple of sales guys could surely handle it... And it seemed Harri had proved the theorem right, with his Radiolinja deal signed just a month ago.

Even with such a miniscule sales force, the employees were filling our small office to the max. Many of our guys

– mainly the original system software team – did still camp at the Microcell office, they simply could not fit into ours. MyOrigo needed a new place.

We had spent about half a year in a standard 15-person rented office space cubicle after lurking around at Microcell's building for a few months in the very beginning. Now, MyOrigo's new headquarters near the bank of Oulu river was ready to move into. It was an old army laundry building, a meter-thick concrete walled monster, ready for strategic nuclear war, it seemed. The building had been in a depressing condition internally – it's easy to imagine what half a decade of army laundry operations does to the inside of a building. The better half of the year 2001 had been spent in remodelling it to be suitable to MyOrigo's needs.

I had a hand in the new office's design process, through a lucky personal compatibility with the architect in charge. J-P had suggested that his long-time architect friend, Kari Kylmänen, would re-design the laundry house to be suitable for a futuristic high-tech company. Kari was game right from the first moment, arriving in leather soled shoes from Helsinki to Oulu's -25 Celsius deep freeze weather for our first meet & greet. I picked Kari up from the airport with Tapsa chauffeuring the company's Citroen XM. Two hours of discussion at the location-to-be, the old laundry house, made me sure that Kari was the man for the job. He knew how to design large spaces effectively with meagre costs per square meter, still making it look cool and expensive. His method was to drop an expensive element here and there to catch the eye, and fill the rest of the space with low-cost industrial materials. I decided then, that my future office room would be one of those expensive elements Kari rambled on about so much...

For visitors, the most interesting feature of the building was the EMS-lab. Electromagnetically shielded and full of expensive measuring equipment, it formed a Faraday's cage impenetrable to radio waves up to 5 GHz. That was needed to make the measurements of our future mobile phone's radiating parts and antennae – but it also looked lovely with its copper walls and copper door seals, built under a descending roof on the top floor of the building. The EMS lab also contained advanced ESD-equipment and high quality electronics tools, so it was the place for HW guys to do their actual soldering and prototyping work. HW – as in hardware – means the physical electronics, circuit boards and components, in mobile industry jargon. Counter-intuitively it does not include the mechanics - the plastic parts and the actual physical non-electronic design of the device.

My main idea for my own office was the levitating table. I wanted a table that would have no legs, no support at all from below or above, it would just stick out from the wall as a 2-meter long noble wood square. Inside the wall there was massive support structure made out of torsional heavy-duty steel bars with structural rigidity of train rails, smitten between the ceiling and floor concrete layers. The table structure was also reinforced under its wooden shell – again, high tensile steel bars did the job. But to the naked eye, nothing was visible – a wooden table really was pushing out from the wall without any features that would explain, how it could stay afloat in the air... It was a cool table – the coolest I had ever seen, but that may be the narcissist talking... My room also boasted an impressive set of Mies Van der Rohe stools and a daybed. Probably the 30 square meter office cost as much as the rest of the initial 1000 sqm floor space, but I thought it was prudent and suitable – hey, I was the inventor in charge!

J-P called and raised on my office design style and decided

to make his even more grandiose. He succeeded with flying colours – his office was an 8-meter high, 200-square-meter minimalistic room with only one single table – his own. If Satan had an office on earth, that's probably how it would look – light penetrating from the small windows 5 meters up, near to the ceiling of the room... J-P's office had naturally full wooden floors, a conversational setting of Mies Van De Rohe cubical chairs, and the best video projector money could buy in year 2001. J-P was never a employee of MyOrigo, he was a board member – and doing mainly graphic design work with his trusted right hand employee-woman, Anna-Leena Härkönen. J-P and Anna-Leena worked for J-P's company, which was called – of course – Metsävainio Design Limited. By selling this work as subcontracting to MyOrigo, J-P was able to finance his own company and afford all that neat stuff – and the gargantuan office. He knew how to spend money in style... J-P had an idea that if – and when, surely – myDevice made it big, he would build an exact copy of that office in Venice, spending winters there. The only difference would be – the floors in the Venetian office would be made out of hand-cast Italian tiles, not wood...the moisture, you see...

It might have been the buzz of moving to the new office, specially built for us in the ruins of the Army laundry, that made me finally feel we were succeeding. But certainly a bigger personal element for me was the first time I got mini-rich, right about the same time.

When I joined Microcell before founding MyOrigo, Jyrki had sold me a generous lump of 100.000 shares of his company. I was able to cash out some 20.000 shares of those – and the buyer was none other than J-P! He believed that Microcell shares would raise their value tenfold in the next few years after listening to Jyrki's sales pitch – which was no surprise, Jyrki could convince a flaming feminist to purchase a Lara Croft action figure. I needed money to support my lavish lifestyle, so I exchanged part of my share portfolio to about 100.000 Euros of cool cash. Taxes paid, I still had much more than ever before. Where did J-P get the money to buy those shares? I never asked – I assumed he had family funds of sorts.

I had sometimes imagined how it would the everyday life be if I was rich. A hundred thousand Euros – "a grown-up hundred" - was not much by the billionaire terms, but for me it gave the impression of what having loose money actually meant. I spent that 100K in less than half a year, mainly using it on expensive dinners, female specimens and vanities. And bough a few old, classic Citroens, of course... Money was no object and I felt no need to save any of it – I was sure, also ensured by Jyrki's words, that in the following years I'd be really wealthy, at least in the range on tens of millions, if not hundreds... My cut of the MyOrigo would be worth a lot if we made it into the market with totally new kind of product. This all gave me a nice, oblivious feeling about money, meaning I could spend that 100K like a first taste of a good drink – swift and enjoyable, as there would be more to come.

So how did it feel to get "rich"? I was 26 years old at the time, and never had any family wealth, so I can say I came to it with open eyes, my mind empty of any preceding experiences. The closest comparison I can find is – unsurprisingly – related to drinking. The first weeks felt like the first high from a good calvados, in the company of friends. That first bite of cold beer, after a hard day at work, like Hemingway so elegantly put it... That's how it felt to understand that I was loaded. The greatest emotional satisfaction I got from being able to pay back a small sum I had owed my father for a long time – as a young lad, almost a decade before, I 'd crashed his Ford

Orion and had never been able to pay back the damages. Until now... He was unsuspecting victim of my new rich-trials, unknowing that I had the cash now after years of meagre living. Visiting my parent's home one weekend, I casually handed him the lump sum in neatly stacked bank notes, stating "now we don't have to speak about money, ever again...".

The restaurant evenings got more and more expensive, at that time I was one of the highest-spending figures in Oulu's quite conservative and small scale fine dining scene. A thousand euros in an evening was not rare, two thousand per night was also plausible often enough. This kind of lifestyle attracts a lot of fair wind sailors, 'friends', who are always willing to party when everything is paid for. I wasn't naive to the max at least, I understood that these 'friends' would vanish if my splurging would come to an end. But as an antisocial fellow, I still enjoyed being popular, even if most of it was fake, an illusion built by the power of flying bank notes. Anyway, what is real in this life... As Hugh Hefner said to an interviewer who wondered didn't it feel bad that those model-girls were only attracted to him because of his money: "What does it matter why the girls are here, as long they are here."

Thus I found out there is one more thing I was really good at. Spending. A pile of hundred thousand Euros had vanished in less than half a year, and all of it was spent well, never to be recovered!

16.

2001 into 2002 – the dawn of myDevice

The second half of 2001 leading into first months of 2002 was one of the most prominent periods of my life. It even seemed to top my previously-unsurpassed six months, the full-time inventing months, in 2000-2001. Loaded with cash from those Microcell stock sales, a new headquarters for MyOrigo with truly esoteric offices for J-P and me – and finally, just before the year broke, the first functional prototypes of myDevice. The protos were initially covered with simple test mechanics, rectangular green machined plastic boxes that housed the electronics. Didn't look like a phone much, more like a military item.

The final shape and size of the myDevice had already been designed some months before. The initial idea of the look of the device was – after some design trials – simple and elegant. It would be a slender thing with ever so slightly curvaceous sides, convex but minimalistic shapes all around. Some wooden models of it were built at great cost – machining, painting and finishing a single wooden mock-up phone cost almost 30.000 Euros, since we wanted the best quality available. These models would be shown to the potential buyers, before actual functional myDevice was available.

Early wooden models were optimistically small in size, in no small part to motivate the hardware engineers to actually place their components in the most efficient way, so that size could be achieved. Maybe they were motivated too little, or the components of the time were simply too big, but from the calculated volume and measurements of the wooden model the final functional myDevice was 40% fatter in volume, and 30% thicker. Extra thickness was not only the fault of the HW team – because of early functional problems with HaptiTouch and the necessary capability of the device to take hits and drops to the floor (back then phones actually had to survive multiple 180 cm falls onto a concrete floor, we believed customers would be returning them to the shops and requesting guarantees otherwise), we had to leave a bigger margin than originally planned for between the actual display of the device and the transparent HaptiTouch plastic panel that was covering it.

HaptiTouch final configuration was based on four force sensors - I had sourly learned from MOP3 that the 3-sensor configuration would never work well enough. Apart from its obvious cost and design benefits, the 3-sensor system just was not stabile enough. When pressing with more force to the corners of the touch panel than the pre-tension (pre-loaded spring force pressing the plastic panel against the sensors when panel was not touched) provided, the calculated touch location became erratic, resulting in unintended or wrong button presses. The lifting force at the sensor became larger than the force holding the touch

panel in contact with sensor, and so the sensor stopped providing valid data. The spring tension could not be increased indefinitely because the sensors could only take so much, so inevitably there would be problems, no matter

99

what we did. So we had to proceed with a 4-sensor configuration, that would practically never experience the lift at any sensor, only the press.

The 4-sensor system had its own issues, however; the most pressing problem, quite literally in this case, was still the pre-tensioning of the sensors, but from another viewpoint. If one of the four sensors did not have any pre-tension and hence would not touch the plastic plate at all, location determination would become impossible. All four of the sensors had to be in contact with the touch panel all the time, loaded with some pre-tension even when the panel was not touched. And that was easier said than done - mass-produced plastic parts always have some tolerance in them, meaning the measurements of each part can vary a fraction of a millimetre depending on numerous production factors, and the design has to tolerate it without becoming unusable. However, the force sensors would not tolerate it, if this kind of variation led to the touch panel not being in contact with all the sensors, all of the time. Since the sensors could only accurately measure forces between 0.1 and 5 newtons (10-500 grams of touch "weight"), we could not simply make the touch panel press harder against all the sensors to ensure that way each of them would be pre-tensioned - it could lead to a situation where one sensor would have too much pre-tension, and would not provide correct force data or would even break down. This problem is easier to understand with a metaphor - if there is a 3-legged terrace table set on uneven cobblestones on the street, all the table's legs will always be on the ground, with about equal force. But if the table is 4-legged, one leg almost always remains in air, and table starts to tilt when its centre of gravity moves. That's exactly the same problem we had to solve - the touch panel was the 4-legged table, and the force sensors

were the cobblestones that the table legs - ALL the table legs - had to touch all the time. Finally, this mechanical riddle was solved by an ingenious mechanics engineer Jukka-Pekka Kemppainen. He designed the touch panel transparent plastic plate and its extrusions -

legs - touching the force sensors in such a way, that it was rigid when linear force - like touch - was applied, but flexible when torsional force was applied. Torsional flexibility ensured that even with a small pretension on the preloading springs, they would provide enough force to twist the touch panel to touch all the force sensors appropriately. And the system worked like a charm from that point on - things left to improve for the future versions of HaptiTouch were mainly the sensitivity (this one required about 1 Newton - or 100 grams - of finger press to reliably calculate touch location) and the tolerance to side forces in a situation when a finger was pushed against the panel at a steep angle, not directly from above.

The sensor element itself, a plastic-cased small electronic component measuring about 7 x 5 x 3 millimetres, was a challenge too. Honeywell's original micromechanical force sensors were too expensive for any consumer mass product - at almost 20 dollars apiece, and we needed four - and they had also mild hysteresis and other functional problems. I had just the guys to solve the problem - my special micromechanical team I had employed mainly to develop us a new HaptiTouch sensor. The dynamic trio - Hannu Moilanen, Janne Remes (another cousin of mine and Timo's brother) and Ville Kampman (nowadays the CTO of Polar Electronics, world's leading heart rate monitor manufacturer) - studied Honeywell's sensor in great detail, improved the elements that needed improving, and finally worked together with Japanese manufacturer HDK (Hokuriku Electric Industry) to create a new mass producible, superior performance force sensor that would cost about a dollar per piece in big volumes (quantities larger than 1,000,000 pieces...). The HDK sensor was called pompously "HDK Micro-Force Sensor with Ultra-Miniature size and High Sensitivity". In this case it was not just marketing talk, but all those virtuous things really were present in that sensor. It looked physically quite a lot like Honeywell's sensors, but inside, it was different. Similar enough, however, that in early prototypes of myDevice we could actually use the Honeywell sensors before the HDK ones were available.

The industrial designers - the shape artists, guys actually defining the detailed outside form of the myDevice, had originally wanted to utilize the unconventional possibility to make a curved touch surface – HaptiTouch was almost agnostic to the shape of the plastic touch panel, so curvatures on it would be no problem. Some non-functional models of curved transparent touch panels were made, but it was obvious at first sight that that the kind of curvatures would capture and reflect all the light sources in the room, so the display would become barely visible. Industrial designers accepted (grudgingly) to use a flat touch panel, only using the curved form in areas not directly on top of the display.

The final myDevice was bloated compared to the wishful wooden models of it – it measured 128x58x23 millimetres, weighting 170 grams. Yes, it was bulky compared to today's iPhone, but back then it was still the smallest of the small, when comparing to PDAs for example.

I thought it was beautiful. It reminded me of 1930's Art Deco designs with its convex appeal. But my opinion might have been biased – this was the first time I was able to physically touch – and try – my own invention developed for the mass markets.

There was also something on the software side to play with – the initial version of the myBook user interface with its WOW-inducing page swipe and auto-rotation, and most of the underlying architecture was there. It was buggy as hell, many features were missing, and still slow, but I was not worried: SW was my turf – I knew we would make it better and closer to a complete package in the following months before the big launch was due. Our original deal with Radiolinja contained a schedule, of course – and before the summer of 2002, myDevice was supposed to be on their shop's shelves. Otherwise, we'd have to pay back that hefty prepayment we got the previous summer. It was money we had spent already, of course...

One major piece of SW was still missing – the web browser. Pretty much everything else in the device we could implement ourselves, but not the browser. Even today in 2014, Mozilla (Firefox), Microsoft (Internet Explorer) and Google (with their Chrome browser) compete about which one renders the web pages with most compliance, and loads them fastest. Back in the beginning of 2002, the situation was much worse... compatibility problems on many web pages, even when browsing with Windows PC, were numerous. It was common to find a notifier on a web page stating, for example, that "This page was optimized for Netscape" ... And we had to make if not all, most of the web pages work with our myDevice – otherwise what would be the point of my revolutionary Virtual Mirror tech?

Since Java was a growing buzzword in those days, there were already a few Java-based web browsers in the market, offered for licensing. We were looking for wide compatibility, the capability to render on big screens to suit our virtual screen tech, a high optimization level and low hardware performance requirements to suit our low-power-consumption mobile processor. After some checking, we found out that the best match for our needs was a browser designed mainly for TV Set-top – boxes, called Espial Escape. It was a stroke of luck to find a browser as good as that compatible with our Java-based application platform. It rendered most of the web pages of 2002 surprisingly well, almost on par with real PC web browsers like MS's Internet Explorer and Netscape.

The integration work – ergo making it run in our system -

of Espial's flagship browser was not trivial, though. Already back then there were so many variants on Java that it was hard to keep count of them. There was Java 1.2, Java SE, Java ME for "normal" mobiles, Java EE for servers, Microsoft's own J++ „'Java', and so forth. myDevice utilized a version of PJava, which was something in between the full desktop versions and mobile editions of Java. There were hundreds of man-hours of work before we even got Escape to compile to our specific version of Java platform. So much for general Java compatibility... The first time we actually got Espial's browser running in myDevice was cause for a small celebration and champagne. One of the biggest risks in MyOrigo's software development was now minimized, with 'just' the numerous remaining issues on compatibility between our platform and Escape to deal with.

As was usual in our software department, David had been in the central role pushing through the final crunch, hand-coding and modifying Escape's internals almost hacker style to actually get it humping in our steam roller. But this one he couldn't have done alone, four software guys at MyOrigo were assigned solely to the Web Browser Experience development. We fully understood the meaning of a truly seamlessly working web browser, it was the most important and critical application - maybe apart from the actual phone dial app – in our device. And so – later after many thousands of man-hours of work the web browser worked as well as anybody could ever expect; albeit much later than I originally planned...

17.

I Code

Despite all my technical successes and thrills during the beginning of 2002, I felt I was slipping farther and farther away from my first childhood love - coding. Yes, I'd designed most of the critical architecture pieces of SW in 2001, with Jari and David mainly on the receiving side. Yes, I had penned the myBook specification and features – and its implications to apps. I was the original architect of HaptiTouch software algorithms. I had certainly done my job as the inventor and CTO. I had even gotten into habit of making those god-forsaken PowerPoints, to present our revolution before it actually existed. In all measurable terms, I had delivered!

But I didn't code any more, not for a year already. That spec and architecture definition work, even advising on algorithmic design - that was all like a hazy cloud compared to a hard rock of real coding, when it came to my definition of work. One element of me seems to slowly shut down if I don't actually code something myself, and I get dumber every day – sliding inevitably closer and closer to those grey middle managers or self-righteous VPs.

I had to do something about that, if I were to remain sane,

remain a whole man. I decided to start to code again. It was easier said than done – I was the CTO and nobody expected me to touch the actual code. Except David, who seemed to understand my pain as a fellow coder – he advised me to get to work, solve something, anything, that was not properly working yet.

Despite the doubting looks of some of our software engineers ("why does he start to micromanage, fiddle with our code? ") I installed the complete myDevice software development environment onto my new, powerful PC that was integrated inside my floating table. Finally, there would be some good use for my two Silicon Graphics displays, which were being mainly used for PowerPoint creation and email reading thus far. Using that computing power on trivial tasks of those sorts, it was like going to a mall to buy family groceries in a Ferrari F40 – inappropriate. Now my gear would get to real work, stretch some hardware muscles in running Tao's complex devenv with MyOrigo-specific numerous extensions!

The current version of Virtual Mirror in myDevice was, bluntly put, hilariously bad. MOP2 and even MOP3 had Virtual Mirror running at least to some presentable extent – but in myDevice, the virtual screen jerked around like in a bad hangover, juddering in 20-pixel jumps instead of smooth, fluid motion that was essential to good hand-eye coordination. The user had to feel he was actually using a real hand mirror, and that was not the illusion available now. So I decided to recode Virtual Mirror myself.

what is the basic modus operandi of a hand
mirror? When you tilt it, the image you see
through the mirror changes, pans around. In our
final myDevice mass product, the only motion
sensor was 2-axial accelerometer. Back in the
day, there were no single-chip 3-axials, and
adding a second sensor to detect the third, z-
axis in right angles to the printed circuit
board, would be too costly. And not necessary
by my calculation - two axes would be enough.
But how can you tell the device's orientation
from acceleration? Gravity is the answer. As
there is constant acceleration on the surface
of Earth because of gravity, 9.81 m/s2 to be
precise, the direction of the gravity vector
can be determined with acceleration. When the
device is at rest - or at a motion slow enough
- we know that there is roughly 1G of
acceleration affecting the device. Now we had
one axis missing since we had only a 2-axial
accelerometer, so things were little bit more
complicated. 2-axial vector length was less or
equal to 1G, the rest of the 1G had to be on
the third axis. And calculating the amount of
that was simple: 3rd axis acc $zAcc = 1 -
sqrt(xAcc*xAcc + yAcc*yAcc)$, where x and y are
the respective acceleration axes that sensor
did provide the data for. So after this, the
direction of the gravity vector is known. Of
course the gravity vector always points to the
centre of Earth, and knowing that, calculating
the device attitude towards the Earth from that
vector direction was straightforward.

Problems arose when the user was at the same
time moving the device - inadvertently, for
example riding a bus - and tilting it, trying
to use the Virtual Mirror. The total
acceleration the device experienced would often
be more or less than 1G - there would be 1G
acceleration from gravity and something added
(or subtracted) by the movements the device
did. So in a practical use case, total
acceleration would be anything from 0.5 - 1.5
G's, making it impossible to separate the
gravity vector from user-inflected motion
acceleration of the device. Practically, using
it on a bus might feel like holding a magnifying
mirror - small movements of the bus and hence
the user would result in jumps and inadvertent

pans on the screen.

The other issue was the user's hand itself. Accelerometers back then were already quite good regarding accuracy and noise, their resolution was about a thousand of a G (1 milliG), and noise was not many milliG's too. But the user hand, even when very stabile, shakes and wobbles with figures hundred-fold greater than that. In practice, the user's hand could result in noise of 0.2-0.3g even when sitting down on a chair at home. That human noise had to be filtered out – that's why the current SW was so jerky, the boys didn't know how to filter out the noise and leave the fine-grained, wanted tilt signal intact, so they filtered out too much leading to a retarded user experience.

I made the decision not even to try to separate the gravity vector from the general acceleration of the device. It would be futile – like sitting inside a opaque ball that is rotated in giant's hand, and trying to guess where exactly the ground is, and which part of the acceleration is coming from the actual motions of the ball. It would be impossible even to a human, much more so to the mobile CPU's of 2000s. So, the user would have to use his own hand-eye coordination to balance out the wobbling when panning the screen on a bus or a train.

Filtering out the user hand tremble and wobbling – and at the same time keeping the snappy response, smooth operation and fluid motion of panning intact - was a challenge. In addition, I wanted the filtering to minimize the effect of gravity and motion acceleration vectors mixing in real-life use cases. We sat down with one of the best math guys in my company, Manne Hannula, and designed a multi-stage adaptive filter that combined marginal hysteresis, an infinite impulse response filter (IIR) and a finite response filter (FIR). FIR was just a sliding window averaging filter for 16 samples of incoming data, making it practically a quarter-second moving centre averager for the motions. IIR was simple design of relative balancing filter – only 5-50% of the filtered final data sample was taken from the new data sample, the rest of it taken from the previous filtered final data sample.

Hysteresis prevention meant in practice, that before the display actually showed any motion to the opposite side of the previous direction of the motion, there would have to be an 8-pixel difference in the calculated position. This all had to be coded in C (low level libraries, what the Virtual Mirror driver quite naturally was, were most conveniently coded in C for Tao's Intent OS, the only higher-level code-like Application framework and the apps themselves were meant to be coded in Java). I did the coding of the initial version in less than an hour. Then the fine-tuning of the filtering started – and it took longer. We rotated the devices in our hands for hours and hours, watching the response on the screen, going back to previous parameters and putting in new ones as we went. Each small change – which there were hundreds of after all the tinkering was done - required a new version of software to be downloaded via USB cable to the myDevice we used for testing. One minute to do that each time doesn't sound like a lot, but when that is repeated hundreds of times, the end result was that hours of the total development time was spent on waiting for the new software version to be compiled, flashed and started on the device.

Tuning of the filters was not a forward-only procedure. We could work for an hour, and realize that our currently tuned version was worse than the one we had an hour ago – and defaulted back to the previous version. Everything affected everything, it was a little bit like building a house of cards – if you add or change something, you had to keep in mind the effect it might have to the other elements in the system.

But we got the filtering tuned and done, finally.

One inherent limitation of the system was the inevitable size of horizontal panning when the device was held in 90-degree right angle to the surface of the Earth. Then the gravity vector points directly along the axis of horizontal rotation, and hence does not move at all even when device is rotated around that axis. So in that specific use position, the horizontal tilt would go unrecognized – and close to that position, the resolution of the tilt calculations were worse. Luckily, in normal use cases, a mobile phone is almost always held in quite a shallow angle towards gravity, so this was in actual use not a problem many would even notice.

I was hopeful that I could achieve the zooming function by detecting the motion of the device farther from and towards to the user. It would involve a totally different kind of mathematics – namely, integrating acceleration other than gravity first to speed, then taking a second integral into a location – and using dynamic changes in it to zoom the virtual screen in and out. I failed, of course, the 2-axial accelerometer was just not enough to provide the data needed. And actually, even 3-axial wouldn't cut to the chase – but that would become obvious to me only years later, with a very different device development project... So the zoom had to work by tilt too. We had two 2-action side buttons, one was used for enabling the Virtual Mirror – freeing it to motion, so to speak – when lightly pressed, the other button I decided to reserve for zooming. A light press of the left button would enable zoom-by-tilt, a hard press would toggle between different, pre-set zoom levels. It worked OK, but was nothing spectacular compared to the Virtual Mirror motion control.

The cursor was the last task. I believed that a mouse-cursor-like arrow on the screen would be needed, so the user could point and click web page buttons without touching the screen physically – and the cursor would be much more precise than tip of the finger anyway. I implemented my nice idea to move the cursor – it would move in synch with the Virtual Mirror motion, only in the opposite direction. So if the user panned the screen to the upper left corner – moving the screen to the lower right doing that – the cursor would travel to the upper left corner also. It was a very intuitive way to use the cursor, and many users preferred using it to click buttons and other stuff on web pages rather than their fingers. An added bonus was that the absolute location of the cursor on screen also showed where on the virtual screen the user currently was. If the cursor was in the centre of the screen, it meant the virtual screen was also centralized that

moment. If it was a quarter towards the left
edge of the screen, the Virtual Screen was also
a quarter towards its left-hand boundary. Easy
peacy lemon squeezy.

After some days of intensively laborious testing and fine-
tuning, my algo was running great, smoother than
anything on the previous protos. I measured – panning
and zooming large test images on the virtual screen – the
screen refresh rate was more than 30 frames per second,
meaning almost buttery-smooth motion on the screen. To
my great satisfaction, it was really giving the illusion of a
large physical display floating in the air in front of the
user's face. Inventor's Delight filled my veins, I did a small
tap-dance around my extravagant table.

I gave that piece of code to Jari, notifying him it might not
compile directly as-is in the latest version of current
myDevice codebase. He stated discreetly and simply: "If
that happens, I'll make it compile" and strolled on to
continue his work.

So myDevice got its Virtual Mirror by my immor...., ehm,
mortal coding hand, and it was the best WOW-effect we
could provide. The thing was sure to sell like hot cakes
now, I thought...

Even more importantly, I felt like myself again. I was a
whole man again. Not only a CTO and a day-dreaming
inventor – but a master coder also. Prinssi Eversti.

18.
Patenting – a practical joke played on all private inventors

The Press and the general audience perceive patents as protection to the inventor. Patent your thing and money starts flowing in almost automatically, right? The first question when you say your tech company was bankrupted, often is "didn't you patent your ideas, you could have sold them?"

Yeah. I am the principal inventor in some 100 granted patents, and never received a single dime from one of them personally. My companies did get some small amounts, but it was never the right time to extract money from my firm, as the investors constantly pointed out. The right time would be the actual IPO or the trade sale to some large company...

Patenting, to a small-time inventor, is more like a bad joke. An expensive, bad joke played by the preconceptions, patent advocates, and investors: you should patent your ideas, otherwise you will not be protected from those big, stealing companies... Reality flash: you will be unprotected no matter what you do.

First; big companies file an uncanny number of patent applications each year. IBM filed well over 30.000 pat. apps in 2010. Over 30.000! How can they have so many new meaningful inventions in a year? Well, they can't: the idea is just to file patents, just to be sure, even for ideas that will most likely never be implemented. And there are 1000 guys who's pay check depends on the number of patents filed, so it's a PMM, a perpetual motion machine that keeps spinning. This means that whatever you try to patent, there are bound to be at least 100 granted patents – usually owned by those giant companies – in the same field as your invention. You might be surprised – you never saw anything like that in the market. But industry giants implement less than 10% of what they patent!

Second; if you have been able to weasel your patent in an empty hole of protection space left by those innumerable amount of patents granted each year to the fortune 500 companies, your patent is almost surely not valuable – it is too specific. To get your patent granted you had to shrink its protection scope small enough for it to be novel compared to those millions of patents already granted. So in the end, it protects almost nothing – modifying one single insignificant detail about your invention can result in a copy that does not infringe that patent.

Third; even if against all odds, by a blind strike of unbelievable luck, you have been granted a meaningful large-scope patent that some big company infringes, there is no money in it for you for a long time – if ever. As patent law states, there is no official or authority actually monitoring who infringes your patents – you have to do it yourself. And, you have to prosecute that infringing party via an extremely complex legal process, including court hearings and all the bells and whistles, to actually get them to pay anything for that infringement. If your patent was not granted in USA, forget it – the possible monetary gain is so small that the costs of going to court-dance with mammoth will be impossible to justify – just like playing 110% or nothing with 50/50 chances to win. Nobody finances it! If your patent was granted in USA, you can maybe find some investors or big patent law offices who will sue that mammoth for a percentage of the possible gains. Unfortunately, big companies know the drill, and have a small army of patent lawyers on their bankroll to handle situations like this. They will throw everything possible – and the kitchen sink – against you in form of tens of thousands of pages of documentation, laboratory notebooks, prior art findings from Japan that patent officials who granted you the patent never found, et cetera. Court cases take many years, and can take decades. And this only applies to absolutely the most promising cases that are clear-cut infringement cases.

If you finally get some money from your patent, you can consider yourself luckier than a lottery winner. Patenting is the all-time hoax played on us, the private inventors with meagre resources!

How did this happen? Wasn't patenting meant to protect the small inventor, not giving fortune 500 companies the world-wide exclusivity to invent and produce anything novel? The whole concept of the patent – which should protect a physical invention, physical device – was lost in 1870's when the necessity to actually present a working device to the patent office was dropped. From there on, the big companies have utilized this loophole to the max – less than 10% of the ideas patented have ever been implemented, brought to reality. What does it mean? It means that the possibility to actually implement something new is almost always blocked, if you comb the patent databases – there is bound to be a granted patent about something like it, and that patent is usually granted to some Fortune 500 company.

Patenting is a numbers game – an extremely expensive numbers game for Mammoths. Show me one case where a private inventor got any net gains from patenting, I'll show you thousands where they didn't.

Why do the investors of small companies then still require that significant part of the funds be used to patenting? Well, many of them still think it's the 1990's when small companies still had some chance against the behemoths. The number of granted patents has quadrupled since then, largely thanks to those mammoths... But there is one element in play that must not be disregarded. If your company is acquired by some business entity via a trade sale, it is more than likely that the buyer will require you to have some patents around your stuff, even if in all likelihood they would be valueless. It's because the buyers have investors too, and investor's just like those patents so much...

And finally, at least there is something good in patents: you can stamp your product with the famous statement "Patented" which gives it some charisma, supposed meaning that there is something really novel in it. And if your business remains small enough, no big company will be interested in suing you or even studying your patents, even if you infringe theirs.

So what does all this mean to the standard inventor, who just got his first 100K as seed funding into his new company? Patenting should only be considered as a marketing measurement against somebody tipping over your business from the start with the classic "they don't even have any patents..."– but there are almost never big gains from patents, just costs. Let the investors bear those costs...

When starting MyOrigo in September 2000, I had no idea what patenting meant in practice. I was under the general influence that if I have an invention, I just apply for one patent and I am protected for life, nobody can copy my tech! I didn't know then what I know now.

MyOrigo's patenting process was quite enthusiastic at first – we had sufficient funds in the beginning via Jyrki's generous hand, and nobody in our team – including Jyrki – understood the true grim nature of what patenting means to a small company. So we started to file for patents at a growing pace, finally ending up with 19 patent families, and well over 60 granted patents worldwide from those families. A patent family means a line of patent originated from a single patenting idea, modified according to local regulations – and language - to each continent, and continued applications filed later.

We filed a patent application for the Virtual Mirror. We filed it for the HaptiTouch. We filed patent also for the original motion control idea, the hole-in-a-cardboard - trick. We used the best patent offices in Finland we could employ, and filed globally – in the EU via the EPO (European Patent Office), USA, China, Brazil, Russia, the works. In 2003 we started to get our first patents actually granted. After four years of operation in 2004, we had all of those 19 original patent applications exponentially sprout all over the world in different patenting procedures, many of them granted, even more of them still pending and sucking money into their process costs. Each patent office - in each country – found their own problems with our patent applications, and each of them had to be answered in their local language, using local patent attorneys as translators and handlers... the costs piled up.

Nokia's hate against Jyrki showed itself even in the course

of patenting – some of Finland's big patent office's we tangoed with informed us after a while that they could not work with us any more – Nokia had blatantly stated that any patent office working with us would not be working with Nokia. And Nokia was the biggest client of the major patent offices in Finland – so we were left using the small ones that didn't work for the Big N.

MyOrigo ended up spending more than one million Euros on patenting in total. It got back a fraction of that when it sold part of its business to F-Origin in 2004. But before that MBO sale in 2004, many things needed to happen.

In 2002, the challenge was not patenting – in our naive view, all was going as planned in that regard. The challenge was – get myDevice ready and certified for sales, and sell hundreds of thousands, preferably millions of myDevices' to some willing customer!

19.
Hero – not!

It all sounds so good, so proficient. I did this and that. Patented those things. Even coded that cool thang. Was I the real-life Iron Man, doing everything just right, steering my company towards inevitable victory and myself towards insurmountable riches?

Hell no. I was doing a banged-up job in so many departments that often I wondered how long could this go on. It seemed to be like an on/off switch in me – I could do some things really well, others not at all. Was I a binary person?

My social instability and sleeping regimes had not improved, to put it mildly. The technical successes had given me such a boost of pompousness that I was all but intolerable to the normal working guy, especially because I could get away with all of my inconsistent behaviour. I hardly ever came into the office before noon.

How does that mix with the things I did before, those deity-level acts of inventor's Heroism, solving the unique technical problems, and defining what a smart phone actually is, for the decades to come?

Well, it doesn't, really, because that's the shiny side of the flip coin. The dark side got even darker when Harri, our CEO for almost a year and half then, was gearing up to leave MyOrigo...

In the springtime of 2002, the MyOrigo corporation had grown into a team of 40+ people. It was more like OurOrigo by then. In the beginning I had wanted to be on top of it all, deliver my magic touch to every department. But it seemed my Midas touch turned some things to shit instead of gold.

The actual myDevice design project was run by Aimo Vainio. He was a by-the-books guy, always fulfilled his responsibilities, kept regular weekly meetings, and watched over the daily work of mobile phone design closely. He reported directly to the CEO, Harri Vatanen. I had a good handle on Harri, he clearly understood what was my importance to the company and product, so in practice I had pretty much control over how things were implemented to myDevice, keeping the dev kosher in relation to my original spec. But Harri had – quite inevitably, as I later found out – ended up in a bitter argument with Jyrki. It seemed those two were never able to agree, what was Harri's correct stock share of MyOrigo, and proper compensation. Harri had learned from his past experiences and wanted to get 100-page contracts signed, whereas Jyrki was more in favour of one-pagers... The gap between them deepened. Harri was ready to leave the company, Jyrki was ready to kick him out, it was only a matter of time before that would happen from one side or the other.

It was obvious that as our team grew and changed, my pet CEO soon to be leaving, my influence over the design process would be minisculisized. I wasn't so naive that I hadn't understand that, but there was little I could do about it. Mechanics and hardware had already been defined so well, that now the engineering work around them was just focused on fixing bugs, making it more mass-producible, and type-approving (you know, those CE-labels and FCC-certifications that any electronic device has to carry before it can be sold to regular customers). So I deliberately left those mainly alone, to be designed by the pros – practically only checking and advising on HaptiTouch mechanical issues every now and then.

The user interface was luckily already so well capped with the myBook, virtual screen, my handcoded Virtual Mirror and general HaptiTouch functionality, that no major deviations from my original ideas was to be expected there. Software in general was an another ticker entirely. The SW team was almost 30 guys strong, there were separate Applications and System software teams. The system SW was going quite well, the original, Johannes-friendly 4-man team had just one person added to it during the dev. But apps team, they were maybe the most stubborn and unprofessional in my view – and most keen also to pick up on my bad habits, it seemed.

David and Jari, apart from the always-loyal System SW team, were my only real tabs on the SW dev arena. After the initial definition phase got done, actually back in 2001, I left practically all the architecture and structural design responsibilities to those guys; they also reported to Aimo, integrating into the more and more traditionally animated mobile phone design project.

Realizing this was going to be like that, even more in the future when most of the recruits were now acquired via an official process, I gradually cultivated a new mental attitude. I would only 'advise' and concentrate on things that were the biggest problems in the usability of the myDevice. Actually, focusing my coding skills to fix the Virtual Mirror algo, was a prominent example of that. But there were less and less such problems I could entertain, as the myDevice was getting more and more ready.

So it went. My practical working days were mainly filled with insignificant email exchanges and PowerPoints, management board meetings where little was decided, dulling me down. I was full of new ideas that did not fit into this first myDevice generation, but we had no real money or resources to really start prototyping and specifying the next generation – we had to sell a million pieces of the current one first. It seemed that once I got the Virtual Mirror working right, there would be no Inventor's Delight -moments in store for me. Now it was all the gradual, slow sliding towards a real mass product, nothing new really added, just evolution towards the acceptable function of all the niggling must-have features and apps.

I was bored. Something had to happen soon, otherwise I'd be turned into the typical, grey VP zombie, wearing colour-matched handkerchiefs and socks. Or drink myself to death with Calvados.

20.
Sell-sell-sell...

I'd had the unbelievable stroke of luck, finding J-P and Jyrki and being able to start MyOrigo with them. Luck is the right word – I had top technical skill and ideas that were nowhere else to be found, but I was a terrible sales man. I couldn't even sell myself, which is a basic requirement for any inventor to get funding. In 1999 when J-P and I first met, I could barely introduce myself without a stutter. But those extraordinary fellows saw through my clumsy social skills, into my inventor's heart and mind. One's life depends on such small details, meeting the right people at just the right time. Pre-determined? Who knows. But there is a small difference in the initial parameters for a guy like me, if I end up in a gutter or as a globally recognized inventor. A Binary person?

Three years later, I was still horrible at the social act of selling – at least to so-called normal people. I never really had to – J-P, Jyrki, later Harri and then the sales team guys did that stuff. Knowing how I behaved, they even sometimes advised me to keep out of sales meetings, that happened at our office with many customers. I could appreciate that – I was more a buyer than a seller.

Even though Jyrki had organized that juicy subcontracting deal between Microcell and us, and Harri had sold non-existing devices to Radiolinja (and had got paid for them!), MyOrigo was tip-toeing with money all the time. The monthly costs were already a six-figure number in Euros, and we never had the money for more than a month ahead. It was hand to mouth, and it was tight. I didn't care much, the only effect that I could really feel was the impossibility of starting the development of the next gen stuff. I am sure our always-worried-looking newly-hired financial manager, CFO Esa Juntunen, got an ulcer from our constant funding plunder – Harri got coeliac disease and haemorrhoids, I heard. I got... bored. The funding issues never really touched me personally, I thought in my extreme naivety and pompousness that I was above them. I had personal cash and the company had been able to pay each of my salaries in full, and the development proceeded. What's to worry about?

Dough got more and more tight during 2002. Esa's face was even more reddish than usual, when he had to constantly explain to the debtors and suppliers, why we would once again pay our bills just a little bit late. We hadn't been able to get in really any independent, real income after that Harri's initial deal. Microcell kept supporting us, but for how long could it, our monthly costs could soon finance a small country!

Jyrki had a plan. He believed that the key to the future was in J-P and me – not these new guys, not with Harri who'd already made himself expendable by contradicting Jyrki too many times. He invited us to Switzerland, his remote side-office, to introduce us to a "man who could turn it all around, help make MyOrigo a real company, independent of Microcell". I wondered who could it be.

Landing in Zürich airport (and my skin smelling strongly of old Kerosene...), first surprise was awaiting J-P and me. There was a finicky but beautiful, slender blond lady waiting for us. She introduced herself as Jaana Kenttälä, back office manager! Back office manager... what could that mean... well, into the Mercedes and towards Zug. Jaana drove, and it was instantly clear this girl was no ordinary secretary. There were commanding moves in her driving style, a fluidity of motion combined with arrogance that suggested at least a semi-managerial mind-set built into her pretty little head. I was sitting in the back seat, naturally, trying to insert a word here and there in the pauses in vivid conversation J-P and Jaana were having in the front. I envied J-P, he would always do small talk with anybody, just like that, sounding natural and interesting, full of decadent but witty charm. I sounded like a... slightly retarded boy from the woods, who was in the city for the first time. And probably the last.

Jyrki greeted us in the foyer of his smallish but expensive-looking, business-style-bauhaus office. Coffee mugs in hands (back then, every Finnish male drank only hetero coffee, meaning traditional filtered black ink) we entered the meeting room, to meet Teemu Vasankari. I didn't make much of him initially, at first glance. A wide smile on his kind but regular face, a short army-style hair, a dozen centimetres shorter than me. An expensive but not very stylish suit – and pocket square to match the tie. Archaic business type, I thought to myself.

After minimalist Finnish-style intros with short handshakes, Teemu started his pitch:

"Johannes, if you only have a single pussy hair in your company, you will go and sell that pussy hair." Jyrki immediately interrupted, with a muffled cough and a statement: "Ahem, Teemu, I believe that in this aspect Johannes is more on the buying side...". That's all that was needed to immediately bond all of us to play for the same team – maybe it was the hypnotic casualty in Jyrki's voice, or the actual turning of the tables like this...

Teemu and I became close friends, probably for life, which remains to be seen. He would later become the closest equal of myself on the business side of things. The rest of the meeting itself was insignificant – Jyrki's main motivation had not really been to discuss anything very important in there, it was more like a wake-up call to me and J-P. We went to have a dinner in the centre of Zug, and already by the evening we were on the plane back to Finland. During the course of the day it had become clear that Jaana worked for Teemu, actually, not Jyrki – they were a team. I wondered what would an office day be like with that kind of legs.

I started to participate more in the sales meetings – actually all the meetings, where some sales skills would be needed. To my amazement, most of the stuff discussed had little or no relation to the actual technical facts behind the proposed deal. It was more like courting and bluffing, always exaggerating the positives, forgetting the negatives. It was, simply put, dishonest. But it was clear that both parties expected that, and knew to always subtract half of the positive, and add twice the negative in their heads.

It did not agree with me. I was always a very bad bullshitter when it came to technical details. I decided to operate differently – make it clear that I say what I mean, and do what I say. It was a bad strategy to sell anything, but I couldn't stomach doing it any other way. It seemed our official sales guys were not having very much success, even with their slick talk – and they did not bother to learn all the technical details of MyOrigo's myDevice to the degree I thought necessary, which pissed me off. The future of MyOrigo was not clear. We needed customers, and soon. Teemu was organizing stuff in the background

125

– in the back office, just like Jaana's title had insinuated – but so far it hadn't turned into any concrete deals. Anyway Teemu was tied to Microcell most of his time, he was actually Jyrki's chief advisor on capitalizing on his company.

Harri, our current CEO, was clearly cutting his ties to MyOrigo, spending very little time in Oulu, but did achieve the largest release of MyOrigo so far: Gigantic Pan-European mobile network, Telefonica, officially published co-operation with MyOrigo, promising to launch our revolutionary internet-enabled phone myDevice to Spanish consumer markets during 2003. But it was still 2002, apart from big words, the Telefonica-deal was more like a letter of intent – they wanted to do something with us, but would not actually bindingly order or pay anything at this stage. So what Harri achieved with them was more like a publicity stunt, which could be important too, but... from where would we get actual money?

21.
Apple – round one

I was on the second tour in USA in after-spring of 2002. Most of MyOrigo's sales force was pounding on the doors of European mobile manufacturers and network operators, trying get something on top of Harri's Telefonica co-op. Our first device would support only European networks, so US clients were not really on option. It was becoming obvious we couldn't deliver our first customer, Radiolinja, the promised myDevice in the summer 2002 – the device was just not ready enough for sales, not even type approved yet.

But it was ready enough for WOW-inducing demonstrations, phone calls and internet browsing with motion control. That's what I was touting in the states – in technology sales format. We had decided to start to market parts of MyOrigo's myDevice key technologies – like the finger swipe-enabled touch panel, virtual screen and Virtual Mirror, myBook, as separate components the manufacturers could include in their different devices.

I had met Intel and Analog Devices again with my now fully functional and convincing myDevice-prototype. Intel in particular was all the more interested about it – there was an initiative to start to plan to integrate the virtual screen and Virtual Mirror -features for their CPUs, so every mobile device with their chips could have full-scale internet. Big stuff, a revolution not just to myDevice, but to all future mobile phones... But on practical level it was endless technical talks with Intel's engineering force, only occasionally meeting the same junior-VP-level tie neck, it was more probing than carving a deal, obviously.

Now, in the last few days of my trip, I was in Silicon Valley, sitting in Apple's meeting room in Cupertino. Apple was a minor prospect, it was not clear where they could utilize our tech, just a small side-step on my trip. Their MP3 player, the original iPod with its circular dial-operation donut as the UI control method, was like a bad joke to me after I had been using myDevice as my personal (but buggy) phone for some months already. But maybe Apple would make a mobile phone some day? Maybe they'd like my tech and adopt that – or better still, buy myDevice as-is and let us modify it to suit the American mobile networks? But nobody back then thought Apple could be a serious player in mobile phones. Motorola, yes, they were once the leading mobile phone manufacturer in the world before Nokia surpassed them, but Apple... they'd never made a phone, and their clumsy Newton-PDA had been a commercial flop.

Meeting with their team was going as usual, that's how it went with that kind of successful company. There was some NIH (not-invented-here) in the air, but Apple's human interface people were still clearly puzzled by the myDevice they got to handle and try. I answered a few technical questions, most of which were probing the nature of our new kind of finger-useable touch screen, HaptiTouch.

Steve Jobs appeared to the room at the end of the meeting. I didn't really get to introduce myself to him, he just took one of the myDevice in his hand and started flicking myBook's pages forwards and backwards, again and again, commenting shortly that sometimes the swipe recognition did not work correctly – which was true, it was still a prototype... Virtual Mirror, the motion control aspect of the device, didn't seem to interest him much apart from the general technical concept of the virtual screen and the cursor on it. The full QWERTY keyboard, which even at this early prototype stage was very useable, was more interesting – he asked, why we didn't we make it possible to rotate the keyboard to portrait use position? I told that the resolution would be insufficient, not so much the touch screen, but the actual user's finger resolution – in portrait mode the user's fingertip would be twice the size of the soft keys, which in my opinion would probably uncomfortable be and unreliable to use. My answer went uncommented upon...

The meeting was over pretty soon, most of the details forgotten in the sea of similar meetings with other companies – no protos were left with Apple that time, I actually only had two units with me, my main phone, and a back-up unit. I though it made a good impression if somebody actually called me during the meeting. Well, apart from the fact that most places in America networks couldn't be accessed at all with my European-standard myDevice... Anyhow, Apple was just one – and not nearly the most prominent one – prospect our tech was shown to. I was much more excited about the Motorola connection which Jyrki had been able to arrange. But other guys were handling that, the Microcell guys who were also trying to strike a deal with Moto with Microcell's own phone designs...

The most memorable thing about that second American tour was the attack of killer bedbugs in one low-cost motel near Santa Barbara. The red bumps scarred my back for months.

22.
Nokia – the Finnish chapter

Nokia was by the far the best known company in Finland in 2002. It was the world's biggest mobile phone manufacturer by a great margin, the unchallenged leader in its own field. Everybody thought that the hegemony would last forever – hell, part of Finland's government budget was planned on Nokia's taxes and export for years to come. Nokia, Noksu, Big N.

We – MyOrigo - had a strange relationship with Nokia, starting from the fact that Jyrki's departure from there had not been the subtlest one. The court case with Jyrki and Nokia was still ongoing, Nokia claiming hundreds of millions of Euros in damages from Jyrki's actions. J-P and I had also had quite unfavourable experiences from trying to offer our ideas to the Big N in 2000, before finding the godfather Jyrki.

On the other hand, many of our employees came from Nokia, and of course they had friends in there still. And – being Finnish should mean something too, shouldn't it? So we tried to sell our tech and myDevice via many different channels to Nokia. The sales guys, J-P and Harri, too, approached and demoed myDevice from middle-manager level up to senior VPs, knocking on every other door at Nokia. I mostly didn't, I had never worked for Nokia, and so I was uncorrupted by their styles... I believed we would find customers elsewhere. But if Nokia took the bait, that would be great in many ways. The volumes – quantities of production – would be in the millions, not hundreds of thousands.

Of course the deal never came. There was surely top level decision from Jorma Ollila, Nokia's CEO and Chairman of the Board, that no business would be conducted with any company Jyrki had his hands on. And Jyrki owned indirectly more than 70% of MyOrigo...

Anyway, Nokia had a lot of knowledge about what we were doing, not only from the newspaper articles that started to surface during 2002, but also via insider info. I don't suspect Nokia had any actual spies in our company, but friends do talk... The typical Finnish engineer had quite high morals regarding company secrets, but since the manufacturing and design processes included many subcontractors and prototype devices were basically scattered to many geographical locations, Nokia certainly had a close peek at myDevice sometime in 2002.

Finally, I personally attended one memorable sales meeting with Nokia in Oulu – VP level was present from both of our companies, and myDevice, in its later, almost ready 2003 iteration, was presented to Nokia's engineer-minded guys.

I emphasized the user interface aspect of the device as much as I could, admitting that we surely didn't have all the little bells and whistles Nokia's feature-rich top phones had, but we had the usability of full internet, and a totally new kind of user interface on top of it – myBook and the virtual screen/Virtual Mirror.

Nokia's guys were not impressed, and to my surprise, not even really interested. The Virtual Mirror and myBook swipe that were WOW to almost everybody we showed them – even to Motorola – seemed to pass them without much emotion. They were the feature kings – they quickly dismissed our revolutionary user interface as a cheap trick, a "gimmick", that didn't have any real meaning or saleability. "We have seen different kinds of UI gimmicks from many manufacturers – Samsung has multi-coloured numbers when dialling a number, Motorola uses their clamshell designs to start and end a phone call, and so on. They mean nothing, nothing at all – what means something is the features available to the users, and that they are comfortable with the menu system.". Right, then. I continued to argue the virtues of UI that brought full internet pages and a soft, finger-useable QWERTY-keyboard to the normal-sized, monolithic phone. It didn't take – it was time to show that it was not the young guy from the woods (me) who called the shots in this meeting. Nokia's managers dug up their feature specification guy from one of the deeper dungeons of the Nokia building. The spec guy started to – with a monotonous, bland voice – to grill me about, for example, did we support MMS video, Bluetooth, and so on. Of course we didn't – we had to spend our meagre development budget on the UI, our main selling point, not on the features that really didn't matter to the user in our opinion. The Nokia guys laughed – we didn't have full MMS support and we expected to sell our device as a 'smart phone' sometime in 2003-2004! We had to be crazy...

They laughed even harder in 2007, when Apple launched iPhone which also didn't support MMS (Multi Media Messages) at all. In 2009, their laughter seized up when iPhone became the most popular high-end smart phone in the world... In 2013, Nokia was sold to Microsoft at less than 5% of their value before the iPhone. In 2014, Nokia Mobile Phones don't exist anymore... Well, I guess it was worth a laugh. The Finnish economy isn't laughing, though.

On the other hand, you have to understand the Nokia mentality. Once you are big and successful enough – and they were twice the size of any other mobile phone manufacturer in 2003 - it's hard to listen to any outsiders, because you think you know it all. That attitude radiated throughout the whole organization, and made sure that no outside ideas would be welcome – certainly not one whose presenters claimed to know better than them how to make a smartphone. Thus ultimate success inevitably leads to decadence, laziness, corruption and doom. It seems that in these ultimately successful companies, the game-changing outsider ideas are not recognized with their correct importance. And then when the game is changed (as the iPhone's launch did), success can turn into ruins very fast. In Nokia's case it did, for sure.

The last scene of our almost non-existent co-operation with Big N was enacted in 2004 - Nokia's Oulu branch really bought, with real money, some Software Development Kit-enabled myDevices' from us at the hefty price of 40.000€ each - obviously to study our revolutionary Java-based software platform more closely. I don't know who approved the purchase at Nokia - maybe some sensible VP was exercising his own decision powers with this small purchase. It didn't mean anything to me at the time, I was too preoccupied with much more serious stuff than a few SDK units - it was obvious Nokia was just curious, not planning to do real business with us.

Is there ever a realistic chance to push your own product into a major company's portfolio, if they are operating on the same business sector themselves? Why would the engineers and VPs in any large company admit: "Shit, those guys from that half-ass small company made a phone better than we could do - let's take it into our portfolio!". It would be like signing their own resignation papers. It seems the only way something like that can happen is via an über-powerful CEO who doesn't have to listen to his own engineering force - and he must really love the proposed small-company's product and the people behind the small company to disregard his own people. But Nokia's CEO was Jorma Ollila, and he hated Jyrki dearly, so that was that.

23.
Bosses x 3

In late summer of 2002, Harri was leaving the company. After arguing with Jyrki for a long time over his contracts and whatnot, the spirit of co-operation had frozen. Harri was sure he would be fooled again, just as Sonera had done to him – ultimately he could not accept Jyrki's working methods and style. This had been coming for over six months, now was crunch time.

Jyrki had used his current fair-wind-sailor popularity within the mobile industry (excluding Nokia of course) to land a big recruit at MyOrigo. Mats Lindoff, a Swedish fellow that had worked in a VP level position at Ericsson, was to be the next CEO of MyOrigo! Great – but... Harri was leaving, transferring his workload "in a civilized way" to Mats – and I didn't have much influence on Mats yet. Sure it was me who convinced Mats of our tech's superiority during the recruitment process, but he was coming from the traditional mobile phone industry. There was a great risk that after Mats settled in, I'd have even less say about the technical development in the company than before.

There was hope that Mats would also understand the challenges of a smaller company – in between Ericsson and MyOrigo, he'd worked on a small start-up company called C-PEN, making interactive pencils. A neat and innovative thing for sure, but nothing compared to the complexity of an actual mobile phone, of course.

My co-operation with Mats started surprisingly well in late 2002, when Mats actually moved to Oulu to really act as the company's full-time CEO. I had learned from my previous experiences that I needed to treat at least some colleagues like human beings, and I tried to do that with Mats. Mats was quite an intelligent bloke and understood well what I wanted to achieve with myDevice. I didn't want it to be a just a mobile phone to its users, I wanted it to be like a... drug, that users couldn't put down once they started fiddling with its cool UI features. Mats understood and listened to my arguments so well, that I actually got more interested in participating in MyOrigo's everyday business than before. Even management board meeting became, well, slightly more interesting.

Mats was also a hobbyist magician, one of the best in Sweden, actually. He was the chairman of the Swedish magicians' foundation, with the card trick skills to impress even the engineer-minded, sharp-eyed MyOrigo guys. MyOrigo's Xmas party of 2002, held at the biggest (and only) spa in Oulu, was the best party we ever arranged with Mats presenting his magic skills. Later that evening I was trying to do my magic to the company's semi-drunk secretaries, but failed of course... Despite being twice my age, Tapsa didn't fail, however. I watched with envious eyes...

Mats' history was far enough from Nokia – and Jyrki – that he could establish meetings with some top level Nokia fellows. He had an Ericsson flair to him, after all, one of Nokia's most respected competitors at the time. We had a business lunch with Mats and Anssi Vanjoki who was responsible for Nokia's mobile phone development overall. We presented a prototype myDevice to Anssi in some detail. Of course it didn't take, Anssi was clearly thinking that any UI we could present would surely and easily be surpassed by the Big N - and there were most likely constraints imposed upon us from an even higher level than Anssi himself.

Sometimes my awkward social elements still surfaced, even with Mats. Driving in the downtown of a winter-snowy Oulu with Mats, accelerating heavily with my rally-bred four-wheel drive Lancia Delta HF Integrale 16V, I glanced at Mats. His thin hair had dropped over his forehead under brisk acceleration. I asked stridently: "Hey Mats, I didn't query this before for some reason, but how old are you?". He said he was 40, to which I elegantly replied: "I thought you were much older, that's how you look!". The moment those words left my mouth, I understood that for that a civilized Swedish guy that must have been a gross insult. Nothing to do than to wait for his reply: "Yes Johannes, I also thought you were much older than 28", he tried to get back at me. This kind of thing happened to me all the time, I just couldn't control my straight shooting comments – and didn't even want to.

I can only hope Mats wasn't very insulted by my remark after the initial shock – I bet that kind of thing doesn't happen often in Sweden, where pretty much everybody is nice-nice, not like us Finnish animals. However it may be, Mats left our company to become Sony-Ericsson's Group CTO, my counterpart but in much bigger game of course, just over half a year after he really started to work at MyOrigo. Behind the scenes there was financial trouble – during recruiting, Jyrki had obviously promised Mats that he could concentrate on running MyOrigo, funding was taken care of. During his short time at MyOrigo I think he understood that our company swivelled on verge of bankruptcy almost every month, living hand-to-mouth with Microcell's help, and no stable funding in sight – if

Mats himself couldn't organize any. Maybe he couldn't –
or didn't want to – or maybe there was something else
happening behind the scenes too. I think the cardinal
reason for his leave was that the Sony-Ericsson offer was
just too juicy – a chance to be the CTO of one of the best
mobile phone manufacturers in the world, in a stable big
company atmosphere, surely was very tempting offer for
him. And after all, he could return to his home country,
Sweden. I actually liked working with Mats – a rare thing
in itself for me - so I sulked to see him go.

Next in line for the MyOrigo's rapidly rotating boss's chair
was Matti Paasovaara, who was the managing director of
MyOrigo. Matti was Teemu's friend and recruited mainly
by Teemu, with Jyrki adding some slick, convoluted talk
and stock options to the mix. It wasn't clear to me what
Matti was doing in our company - maybe Jyrki already
suspected Mats will be departing soon, following Harri out
of the door, or maybe there was an idea that MyOrigo
needed not just a CEO who handles mainly external
relationships and publicity, but also an internal manager
of the company - a managing director. This kind of set-up
is not rare in bigger companies, but did MyOrigo really
need it, then a company of some 60 people... Well, at least
there was the next boss ready inside the company when
Mats packed his bags.

Matti presented a new sturdiness in running the company.
He had a way with MyOrigo's employees that I learned to
respect, slowly, apart from my initial personal grudge with
him - he was actually the first real people manager we had
had. With outside-world relationships and presentation,
Matti was quite the counter-image of his convincing and
authoritative internal company manager. When visiting
Sofinnova with him - a large investment bank with its
headquarters in Paris - I had to do most of the
presentation and slick talk. And I'm quite lousy at that,

which meant Matti was even worse. He was really no international marketing and sales guy, no - maybe that was the reason he still had the title of Managing Director, not CEO. Well, nobody is good at everything, I thought.

All these changes in the highest chair of the company of course affected the employees. It was obvious that something was wrong in the upper levels of our hierarchy and funding, the average enlistment time for a MyOrigo CEO was less than a year, each new guy bringing new policies and strategies to the table, at the same time trying to clean the trail of the previous one. The board of managers - including me - was sticking to their jobs well, and that helped a bit - but still company's team spirit was down. Everybody continued to work, but the initial enthusiasm had been lost sometime in 2002, now this was "just a job" for most engineers, and clearly a seamy one for some of them. They had forgotten, become blind to the fact that it was a privilege to be working with this kind of world-changing tech. It's hard to blame them - but I do it anyway.

24.
Personalities of an all-star team

myDevice was slowly getting ready to sell in 2003. It was type-approved for sales in European countries, most of the apps were coded and already running, of course the bug list was still massive... But we were getting there.

Who were the guys who made it happen?

The company had done serious recruiting for all of its 3 years of existence, sparing no expense trying to get the best engineers - and to some extent, marketing people - in. This had resulted in a wide variety of superstars working for MyOrigo among the full 70-person staff, starting with the impressive - albeit constantly changing - CEOs.

As a whole, what should be the structure and staffing of a company like MyOrigo, trying to make and sell a new kind of mobile phone, that no customer had actually ordered or specified in the beginning - an own design to be sold by an ODM business model?

The company had now half a dozen departments. Managerial and financial staff, including VPs, secretaries, and general office people not working on design or sales of the myDevice, was a 10-strong highly unmonolithic unit, full of esoteric characters starting with myself.

Then there was Sales & Marketing with its 3 people, graphics and user interface design with 5 people (including J-P as a subcontractor), Hardware and RF with 7 people, Mass Production Management 2 guys, Mechanics 4 guys, Testing 3 people, Technology and IPR 4 people... and Software, 35 people in total! More than half of the company's own and subcontracted resources were working solely on different layers of software, which was understandable - in building our myDevice, most of the work was on the software, it had to finally implement and show on the screen all the hundreds of features that a mass-producible mobile phone must have.

In the beginning all was simple, just me as the top dog, and accompanying me, some trusted software guys known to me before MyOrigo. But it's no surprise that many of later recruits were also long-term friends and relatives of mine.

Tapsa Korhonen and his son Tommi were the elementary day-to-day runners of MyOrigo, as it was already in fall of 2000. Tommi was a young lad when hired, now, working his third year at MyOrigo, he was no longer the IT manager - he was the actual hands-on guy that handled most of the IT at a concrete level. We had a new IT manager, who had the necessary experience in running a multitude of servers and databases, that some of our smart phone technologies were going to utilize. Anyway, during these years, Tommi had gained a lot of social experience and was starting to blend in to the normal company hierarchy and culture. It was good, no matter what happened to MyOrigo, he'd have a place in the industry from here on.

Tapsa, who originally started as our financial manager

lavishly accepting all my suspicious 'representation expenses', was now an office manager. In practice, he was still the secondary soul of the company, managing all those numerous small day-to-day things not really related to the actual design. You know - buying printers and cables, repairing automatic coffee machines broken by the careless engineers, entertaining visitors, presenting the company space to new employees, tip-toeing with secretaries, and so on.

Esa Juntunen, a real banker who'd formerly worked for one of Finland's biggest banks, had stepped to fill the necessary void in the actual business-level knowledge of handling the company's enlarging banking issues and finances. He was an old-school, extremely polite gentleman, always wearing a suit or at least an old boy - type blazer - I don't know if it was Jyrki or Teemu who'd actually recruited him, but unsurprisingly, I was not a major component in the recruitment process. If he didn't get an ulcer from the company's finances that constantly fluctuated between minimum level of survival and certain bankruptcy, he sure did from my 'presentation costs'. No one in the ranks had the balls to tell to me directly to stop spending the company's money so lavishly to entertain important business contacts in the bars and restaurants of Oulu, instead they complained to Esa. I didn't envy him - but self-righteous bastard I was; I also didn't feel much remorse regarding my actions then or later. I just wanted to think that it was Esa's job to handle the finances, however difficult they may have been. Anyway, I felt quite sympathetic towards Esa, mainly because of his old-school gentleman manners and uncanny skills to talk more time out of our creditors.

After Harri became MyOrigo's CEO in spring 2001, there was a huge cache of engineers he knew outside Oulu that were suddenly up for grabs, and, influenced partly by

Harri's reputation, eager to start working for us. David - and through him, Jari - were stellar guys that any tech company would be lucky to have - they were banging our software to the final status of mass product level during 2003. David was clearly stressed but still very much on the top of things, fixing all those innumerable and unnamed problems that nobody did - or nobody else could. There's always that one guy in any company, the secret guy who fixed this and that small-but-impossible-sounding software problem - in our company it was David.

Jari with his stoic style pushing forward with the performance of a caterpillar - not the fastest, but dependable and robust as hell. He wasn't really a major problem solver like David, he was more the knowledgeable guy that everybody depended on, when changing anything relating to the software architecture or Java aspects of the myDevice.

Our first 'real' marketing guys had also started to flow in after Harri's rep. Sven Weil, a 2-meter tall German man with a happy grin always stuck on his face, was the marketing director. He was a friendly guy and trusted by Harri, arranging behind the scenes for example the numerous Telefonica contacts that Harri alone could not handle. He was also intimately knowledgeable about software technology, he even wrote some of the 'official' specification of myDevice for the customers - pissing of some engineers in the process, they didn't want to implement everything Sven was throwing to the paper. Sven, however, left the company shortly after Harri, so he didn't get the see the end of the ballgame.

Johan Granholm, who later became the Sales & Marketing VP for MyOrigo, was a totally different kind of soldier - he stuck around to the very end. He was a bald, sturdy man with a no-bullshit untechnical attitude - his background was on the construction side of the electronics industry, home automation, electronic building supplies, and so forth. He never used myDevice as his daily telephone, not even in the end when it was ready for sale, which in my mind kind of defeated the point of trying to sell it. But he was pushing a lot of work and got a deal done with a South-African mobile operator when MyOrigo most needed some sales, so I couldn't disregard him as a bad sales guy.

The company needed more marketing power in 2003 when it was becoming obvious that ODM deals were extremely hard to come by. One of the recruits I liked the most was done by Johan - a marketing lady called Eija Hämäläinen, who naturally preferred putting her face in the magazine interview rather than a picture of the myDevice she was trying to sell. I didn't mind, mainly because of the aesthetic quality of her face... and the underlying body with its sizeable protrusions. Doesn't it sound misogynist, by today's standards? Apart from the suspiciously badly-defined meaning of that feminist buzzword of 2015, I can state that whole engineering industry is like that - and me more than most. What's the point of writing about the skills of the girls working in our company when we - most of the guys - were looking at their tits anyway? Yes, I liked Eija more for her big-titted slender body and easy sales gal attitude than her actual verbal presentations of the myDevice product. She might have been quite qualified in that too, but my mind was focused elsewhere. Unfortunately, as a socially limited person I never really harassed her - which I should have done of course, just to be sure - or that one secretary with huge knockers and thin waist. But I sure talked to them, and about them, in a style that would be highly inappropriate by any western world standards today, apart from some countries in the old continental Europe maybe. I fear western culture will collapse if men aren't allowed to be men - looking and treating women with ways they have done for thousands of years.

Originally the very thin hardware and mechanics teams were now getting beefy - the leader of them Antti Ylönen, a young HW guy that came aboard when Juhani was still working for us, he became the chief designer of hardware after Jussi left. I liked Antti, he was a tall and goofy-looking friendly fellow, always ready to listen to suggestions (even from me), and with his organized working style he made things run smoothly in the ESD-shielded lab. The RF- radio frequency part design, meaning the feed to antennae and their configuration, was done by Kai Aalonen - and then there was a separate antenna guy who actually designed the physical antenna - you know, those large metal plates inside the myDevice plastics that served the antennae-function.

The mechanics team that in the beginning really consisted only of my cousin Timo Remes - who wasn't even officially working for MyOrigo, he was a Microcell fellow - was now employing 4 guys with different skill sets. Eero Karhunen was an old-school ever-suspicious designer who handled the day-to-day management of the team, then there was my other cousin - Vesa Tsurking - a prototyping manager of the mech team, and finally, couple of gifted mechanical engineers working on the actual CAD.

Side-by-side with the hardware and mechanics was the mass production team, their role was to work with the factories and component manufacturers to make sure that the product could actually be manufactured. Also, they were responsible for the production line testers and all that stuff on the pipe needed before the product would be ready to sell in its cardboard box.

The testing team - as the name implies, their job is to find and document any bugs the device had, and also to report on the general usability of the device as a whole. They mainly consisted of super-organized pedants who really liked to do organized and semi-monotonous work, at least so it seemed to me. With the bug database growing to over one thousand tickets, their 4-strong team was actually working hard to keep up with the corrections and new bugs.

The Tech and IPR team consisted - in addition to me - of quite academic guys, Doctor Manne Hannula who was our math and algorithmic specialist, Doctor Hannu Kauniskangas, who was the human interface and device feature scouting specialist, Heikki Pylkkö - a mathematic mastermind too that concentrated more on the HaptiTouch math, and finally, Jukka Härkönen, our IPR manager. We had such a numerous and widespread patenting and trade marking operation in our heyday, that a dedicated fellow like Jukka was the only way to deal with it consistently. Jukka was also the main contact between MyOrigo and the local and international patent offices' we were using. Ah, and there was Puuha-Pete - he was shovelled into my team when Aimo realized he couldn't get the guy to do any practical work at all. I of course realized the exact same thing little bit later, trying to exhaust him from the company but failed... The worse workers are the ones who stick to their positions hardest, and know their legal rights best.

The graphics and user interface design team was a blend between artists and coders. I was almost always happy with them - they needed no special guidance, and they clearly liked their jobs, making graphic icons and elements for the UI, and integrating them with code. Tomi Huoviala, later recruited by myself, was our chief graphics designer. I worked with him in the same team in the 90s, at the large software company CCC, and it was easy to convince him to join MyOrigo - better pay and a much more interesting job compared to that software giant's procedural and boring ways.

J-P's role and relation in day-to-day work was mainly with the graphics and UI guys. J-P still did actual graphics and UI design work himself to some extent, which was great - I liked that the final device would have the avant-garde touch of J-P in it. Also the myDevice's final physical shape was largely designed by J-P. In addition to himself, Jukka-Pekka employed with his company - Metsävainio Design Ltd. - three people strong at that time. Anna-Leena, the original graphics lady that had already helped with our 1999 pre-MyOrigo proto, was the spirit of J-P's team, leading the actual graphical design of myDevice with Tomi. This was how J-P funded his own company during MyOrigo-years - his board position was not a paid position at MyOrigo, so he used the same subcontracting trick that we used with Microcell to keep afloat.

The sourcing team - well, really just one person in our company: Juha Rytky, who later became my business partner for years to come even after MyOrigo. Juha had been recruited to the company during the spring of 2001 from the world's leading heart monitor manufacturer, Polar Electro OY. He handled innumerable negotiations and meetings, price hacking and order scheduling with all the components in the myDevice. There were almost four hundred separate components. The Bill of Material (the component list of the device) also included special parts manufactured solely for MyOrigo - the display, mechanics parts and mains charger, to name a few. Juha's role would later grow to managing the whole operations of the MyOrigo as Chief Operating Officer, COO. Juha married my sister, Sini after they got acquainted at MyOrigo back in 2002, so this was turning into something of a family business...

Finally, the hefty, 35-strong Software team - it was divided in several sub-divisions.

Starting from the 'bottom': The system software sub-team - those trusted friends of mine - handled all the lower level code (mostly written in C, sometimes assembler) and drivers. Urpo with his friends handled the part of the software so well that it was almost invisible to the managers - it's always the best sign that somebody is doing their job properly! If you don't hear about some team at most of the meetings, look into what they are doing one day and either raise their salaries or fire them as being unnecessary. In our case, these guys certainly were not unnecessary...

Then there was the Software Framework sub-team with Jari leading the efforts. All the underlying high-level code that many applications were using, like the myBook-framework and InfoBus, were coded and designed by those guys. Jari also headed the whole software team as his side-job...

The applications sub-team was in charge of all the end-user apps - all of which were written in Java. They were the largest of the sub-teams with almost 20 guys - and it was no surprise, as finally, the applications, apps, implemented all the features that the end-users actually see and handle. Phone book, calendar, email, camera app, music player, settings... the list was long. This was also the team giving most problems and bugs by far, as was to be expected. The interactions between apps and mode switches were challenging even without advanced framework, and in addition it seemed to me, that the work morale of that team was the lowest of the company. Why? Well, their job was not really the most rewarding - not really crafting architecture or any of the cool stuff, just trying to make work those endless features that smartphone users expect.

A little bit separate from the other sub-teams was the Software User Interface sub-team - they were quite tight with the applications sub-team, but handled most of the general code that actually rendered the visual stuff on the screen. The myBook UI was largely coded by them. Also, they worked tightly with Tomi's graphics and UI design team, with desks located in the same physical room.

And the list continues. But aaaah... Is it really worth to try to describe everybody and all the moving components of such an endeavour? Apart from entertaining everybody who wants to see their own name in a book, there is no point to it. After all, this is not an inclusive task list of how to build such a company. Although this might be how not to build one...

25.
2003

It would be all too easy to tell numerous slick stories about how things rolled forward at MyOrigo during 2003. Much drunken misbehaviour, even more business meetings - and of course, building the myDevice to a ready-and-approved phone for sales. Me even writing a magazine article about mobile Java, for the best known technology magazine in Finland called Prosessori (the processor) - an act which was supposed to cement our position as world leader in mobile java knowledge.

But revisiting all that would be boring. 2003 was just the posthumous ripples of motion started in 2000 and brought to its innovative and cocky peak at 2002.

Sufficed to say that MyOrigo finances went down throughout the whole 2003. myDevice got ready for sales, yes, but no real revolutions were achieved after 2002 - just gradually more and more of a "whole", functional device, that everybody could use as an everyday mobile.

I let the following event characterize the whole of 2003:

We were sitting in MyOrigo's biggest and grandest meeting room, designed in a post-modernistic Avant Garde style. A couple of our VPs and me - nursing a bad hangover - were interviewing a company from the capital,

Helsinki. Helsinki's pride was here in Oulu to offer us subcontracting services. As we were still the hot potato of the Finnish mobile industry, this kind of presentation was not uncommon, I had to endure many every month.

The presenters looked all business-like. Wearing the latest season Hugo Boss suits and matching ties & hankies, they clearly wanted to make a statement about their German-like professionalism and skills. The boss man of the group wore Armani in order to be distinguished from the rest...

They started their presentation with an aggressive pitch about their mobile java solutions, the Armani man doing the talking about how their ideas could actually improve our current java-based design and so on. The presentation shone like gold, obviously made by high-quality artists with Adobe Illustrator (not PowerPoint...) to further convey their vision and ambition. But... already from the first minute, I noticed quite a few things in their prez that didn't match with what I knew. And this wasn't imagination or cockiness (this time) on my part, I had actually tested some of the 'solutions' these guys tried to sell as their unique ideas - and in testing them, I had found their intricate problems that actually prohibited their use in a real software production environment.

First I subtly tried to make them think about what they were presenting, pointing out the problems, but the speaker was on a roll - nothing was going to stop him. He was in some kind of self-induced hypnotic state where he believed he knew all the answers to our so-called issues. Finally, he reached a slide in his pdf-converted Illustrator presentation where I had to stop the charade and blatantly state: "Sorry Henrik, but I know for a fact that the JIT performance factor you have in there is grossly wrong.". I thought about it for a second and decided to continue with a lure: "Did you read the article in Prosessori last month, it actually had a pretty good piece about the JIT versus Native Compiler issue?".

He took the bait, and answered with a slightly raised and challenging voice: "Yes I did - but did you?"

My reply was uttered with a lowish, slow tone: - "Well, it depends on the viewpoint... you see, I wrote that article..."

I think that Armani guy still remembers that meeting. I know I do - only for the reason that it was one of the very few funny and nice moments of 2003.

And then, after a totally dull year filled with pointless meetings and get-togethers, the end of 2003 came. MyOrigo was ready for myDevice sales but there were no real paying customers. CEOs had left the company one after another. There was no money. And I had no money - I had spent all the cash I acquired for my stock so far, and actually had credit card debts up to my ears. This is how fast the tables turn: just a year prior I felt sure the "golden times" would never end.

26.
Jumping ship: IPR transfer to F-Origin

2004 started with grim feelings at MyOrigo headquarters in Oulu. It had become obvious that the potential tier 1 customers were not ready to buy myDevice as-was or with slight "branded" modifications. The NIH - "not invented here" - effect was too strong, all the possible big partners wanted to innovate and spec their top end phones themselves. It didn't seem to matter how good and novel our approach was - it was just not going to sell via an ODM model.

Additionally, support from Microcell was fading. The guru himself, godfather Jyrki Hallikainen, had been able to cash out big with Microcell just a couple months back; with Teemu's help he had sold Microcell to Flextronix for about a hundred million Euros. Jyrki cashed in more than 80m personally, and as a result changed forever as a person. He lost his hypnotic effect on people and many of those uncanny social skills - or maybe he had no need for those any more so he just didn't exercise them. Also, as it looked to me, he'd lost his interest in MyOrigo and our once mutual plan to change the mobile world forever. He was going to enjoy his multi-millionaire life and invest in steady real-estate, hanging out in Luxemburg, Monaco and Switzerland with his new friends, driving AMG

Mercs...

It seemed the only change for MyOrigo was to try to push for smaller deals with smaller partners. South-African mobile operators, Taiwanese small-time mobile phone manufacturers, Chinese local markets with miniscule profits and prospects... Those seemed to be the only channels where myDevice could ever hit the markets - if MyOrigo didn't go bankrupt before then.

I had realised some months back that this was not going well, and that in all likelihood there wouldn't be a pot of gold waiting at the end of this rainbow. The contrast to my high spirits just two years back was immense. I was depressed and detached from my original goals, drinking away my daily frustration.

Just when the things seemed at their darkest, New Hope came from an unexpected direction. Teemu Vasankari, the small funny guy I had met for the first time couple years ago and scarcely ever since, had been monitoring us at the MyOrigo from a distance, but much more closely than I had realized. He had sensed the possibilities in the tech we were building - he was able to hide his enthusiasm quite well, but I believe he already knew in 2003 that MyOrigo was doing something truly world-changing, not just building hype. And now, at the beginning of 2004, he surely realized with his sound financial instincts that the original MyOrigo business model was likely not going to cut it - and Jyrki was not really interested in doing much about it anymore.

Teemu had a radical proposal for me and J-P at the beginning of 2004. With his team of friendly lawyers, we would rescue all the IPR, all the intellectual property and design wisdom from MyOrigo, into a new company that would focus on selling that radical technology and development skills. All the rest - the current myDevice, and almost all of the employees (and costs) would be left to MyOrigo. MyOrigo would continue to try to push myDevice to markets just like before.

The operation would be executed as an MBO, a management-buy-out. In this case the 'management' would be J-P and me, and the stuff we would buy out from MyOrigo would be exclusive rights to all the patents and innovations. Virtual screen, Virtual Mirror, HaptiTouch, myBook, Java framework... all was up for grabs. Jyrki would go along because he'd still get a major cut of that new company too, and frankly, there weren't many other options to try to rescue something from MyOrigo in case it went bankrupt.

We took the chance with J-P. The MBO was done for a nominal sum of 30.000 Euros – the basis for such a low initial compensation for MyOrigo was that it retained all the usage rights for its own designs and products with the deal.

F-Origin OY was founded in spring 2004. I jumped the MyOrigo ship and became the CTO and largest private owner of the new company. A small group of selected people followed me, and we moved to another office near the MyOrigo premises. I had to abandon my floating office table and all those secretaries...

The name I christened the new company - "F-Origin" - was meant to be expanded and explained as "Finnish Origin", bearing also some resemblance to MyOrigo. It was out in the open that our tech had been developed at MyOrigo so there was no need to hide it in the name of the new company. However, that wasn't the whole truth about the interpretation of the name. Now after ten years I can reveal that it meant "Fuck Origin" for me. It characterizes my act of jumping the rotting MyOrigo ship.

According to the MBO deal, I was able to smuggle most of my personal stuff out MyOrigo to F-Origin. That included the Mies Van der Rohe chairs and day bed, Eileen Grey coffee table, my powerful computer with its Silicon Graphics displays, and my working chair that was good enough for Michael Douglas at his London Office in the flick Money Never Sleeps.

Apart from all the evil glares caused by the ship-jumping operation, and the hassle of documenting and executing a full-blown MBO, I was happy and enthusiastic for the first time after those long-gone 2002 Inventor's Delight - moments. There was now new hope, a more flexible management team, with Teemu himself acting as CEO. This would become something different to MyOrigo with its already dragging and too heavy feet - this would be an agile company integrating my most revolutionary innovations in other companies' phones, just the way they wanted it, into their own basic designs. I would get to change the world after all, virtual screen/Virtual Mirror and HaptiTouch would soon be the defining elements in all high-end Motorolas, Nokias and Samsungs!

27.
The Different Child

It was all going to be different at F-Origin. I swore to myself I would not repeat the MyOrigo mistakes.

- A small and agile organization without any overlapping personalities with mine.

- A technology licensing and design business model: forget that cursed ODM that obviously didn't work in high end phones.

- Focus on companies based in the biggest growing economies in the world, USA and Asia - not those old-school European firms.

- The marketing approach would be customer-based, not product-based like at MyOrigo. We were not going to push too much what we had but adapt out offerings to what customers actually wanted to do themselves.

So pretty much everything was to be the opposite of MyOrigo. It couldn't go wrong now! The only common thing was - apart from the "Johannes" factor - the original innovations I made with J-P back in those golden years, 1999-2002. I still believed they held the key to the future.

The new approach obviously necessitated a new marketing and management team. But even more than that, it needed a steady boss that would not be stupefied by J-P's or my grandiose personalities and the unconventional "sci-fi alien tech atmosphere" surrounding the whole thing. The more I got to know Teemu, the more I understood he was the perfect CEO for all my practical purposes - he was the one that didn't want to be one! It was like that utopia of having a tyrant that doesn't crave power but has taken his position reluctantly. Teemu even kept his title as "Acting CEO" to signify this clearly. He just wanted to make things happen and cash out smiling all the way to the bank - he didn't care about having his face in a newspaper or bossing people around. He even tolerated my expressionist behaviour so well that I had to wonder if he was on meds. But no, that was just him - he was able to see his final goals through these momentary clashes and kept his direction clear.

The rest of the management, at the beginning, was really just David Narraway - my trusted SW guru from MyOrigo - and me. J-P was again the invisible hand in the background shadows, providing design input and various graphics elements for our tech, officially manoeuvring as Chairman of the Board.

The marketing team was a bright new mix that included Teemu. Silicon-Valley-well-connected Gloria Maceiko from AT&T was the big gun that was hired to jump start our business. She was an all-American, all-smiling, all-customer-supportive sales person that was quite the opposite of typical Finnish sales guys. And the complete opposite of me, of course - I thought that was just what we needed. They would sell something that the customers actually wanted and at the same time sneak in our proprietary tech with their smooth style. And Teemu himself was the master sales guy, his key role as the CEO would not be managing our miniscule team but finding customers and funding.

The rest of the resources were borrowed from MyOrigo and subcontracted from other IT companies in Oulu, when needs arose. For me this kind of model was convenient, I never had to push my own views, since I got to do the specs with David that subcontracted employees had to meet in order to get paid. No more endless haggles about what should be done and why - and why it should *not* be done my way - that had been the Achilles heel of the team performance at MyOrigo.

And so, we basically started operating in a single, bland room in Oulu with MyOrigo downstairs. There was no spacial glory, no special office design this time - this time all the time and money was focused on getting sales. Gloria was working pretty much on her own in US, with her own connections, and we only saw her on special occasions. And Teemu was also travelling at least two thirds of the time, so our office was pretty empty at times.

My time was mostly spent on organizing and developing the already vast patent portfolio and creating presentations about Virtual Mirror, HaptiTouch, myBook - and their integration into a modern high-end phone. I didn't like it but I understood it was the necessary prelude for the actual deed - getting to insert my tech into a Nokia or Sony-Ericsson phone.

My lifestyle actually improved quite a bit during the beginning of 2004. I was drunk less, and well, not getting fatter all the time. That wasn't really the result of any single new element in my life, just all these changes together seemed to wake up my inner inventor again. But I knew it would not last long if nothing came up. And frankly, we all knew F-Origin had funding for only a few months ahead, mainly injected into the company by Teemu and his lawyer friends to pay our salaries and travelling costs. If no big fish would be caught, if no major company started to utilize our tech - and pay for it too, preferably in advance - F-Origin would go under before the year 2004 ended.

So we were on a shoe-string timetable and the clock was ticking fast.

28.
PlastiTouch

Already whilst inventing the first version of HaptiTouch in 2000 I had wondered whether the steadily developing plastic manufacturing technologies would someday enable inserting HaptiTouch into a single piece of plastic, without any moving or separate parts, into a completely water- and airtight package. In myDevice, the HaptiTouch consisted of several parts and their interconnections - transparent touch surface, frame, force sensors on the circuit board, and multiple gaskets, screws and springs to tie it all together into a single sandwich of a force-measuring, touch-estimating package.

Perlos (nowadays in 2015 Lite-On Mobile Ltd.), one of the leading plastics component manufacturers in the world in 2004, took a serious interest in HaptiTouch. It happened after I promised them (foolhardily as usual, based on my intuition...) it could be integrated into a single piece of plastic, and provided to mobile phone manufacturers as simply as passive displays covering plastics were provided those days.

In the summer of 2004 we constructed the first PlastiTouch prototype at F-Origin. It was based on my knowledge of how such a part should be moulded; some mechanical subcontracting was used for the actual 3D design and then we crafted the rest of it together ourselves. It was pretty revolutionary stuff - again... a single transparent plastic part that acted as a touch panel. The bottom of a standard issue myDevice acted as base for the proto - we figured it was much more convincing attached and working in a real phone, not just hanging in the air by itself.

It had no moving elements, no sealing, completely air- and watertight in the whole touch area and frame. Just like a plastic block that could magically sense where it was touched. It was the touch tech of the future. And with the right manufacturing technology Perlos could provide, being one of the high-tech plastics global leaders, it could be made dirt cheap compared to any existing touch technology. And yes, even more importantly, I got my small Inventor's delight - moment when I made it work the first time...

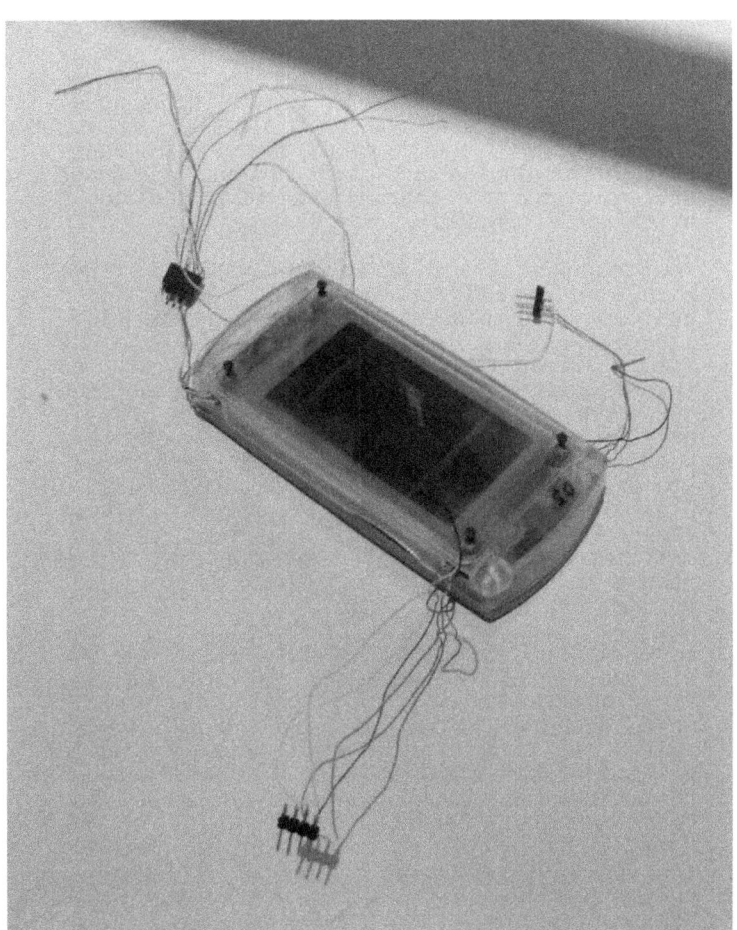

The 2004 PlastiTouch advanced prototype that really kick-started Perlos' interest in F-Origin tech.

The problems of constructing such a single piece of plastic were numerous and tough as a stoned nut. First of all, the structure of the plastic would have to be suited to a cheap injection moulding process, the system commonly used in the plastics industry to mass-produce parts cheaply and in vast quantities. That necessitates surprisingly many limitations of how the actual plastic part must be shaped. It's like sand moulding on steroids - if you cast lead in a sand mould, the mould cannot have tight angles and deep holes, it cannot have protrusions that would make the mould stick and irremovable, and so on.

The shape of the plastic part itself - even with all those injection-moulding limitations applied - had to be such that it would be rigid laterally, but flexible torsionally. The lateral rigidity was necessary to transmit as much as possible of the touch force through the force sensors to the base portion of the plastic part. The torsional flexibility, on another hand, was necessary to allow for imperfections, "tolerances", in parts like the PCB and the PlastiTouch plastic part itself. Those imperfections would inevitably lead to sub-millimetre changes in, for example, the fixing height of each screw attaching the plastic part to the rest of the device.

In addition, although a single plastic piece with no seams, the part had to logically and functionally consist of a relatively rigid frame that would not be touch sensitive, a relatively rigid touch panel portion, and a quite flexible plastic "neck of land" between those two functional portions. In a way the "neck of land" replaced the function of the original HaptiTouch porous gasket seal between the touch panel and the frame of the phone - now it was just all plastic, and seamlessly a part of the single structure.

That was the mechanical part - the beginning. We got it right pretty early on because I knew exactly how it would have to be shaped and constructed based on my 4-year HaptiTouch experience.

The electronic part in this prototype stage consisted of the same HDK micromechanical force sensors we used in myDevice's HaptiTouch. They had special cavities inside the magical monoblock plastic part, with copper wires leading the signals out of them. And for practical reasons, we had to put calibration

screws in there to pre-set the force on the
sensors before a finger was pressed on the touch
panel area.

Obviously in mass production, the whole sensor
system and pre-set force would be handled
without any external wires or screws -
actually, that would be the main design input
Perlos would bring to the project in addition
to the true mass production capability. Perlos
was conveniently just entering the era of
laser-printed copper wires integrated into
plastic parts (for antennae purposes,
originally), and this new plastic technology
would coincide nicely with my plans for
PlastiTouch. The sensor signals would be led
out from the PlastiTouch piece by the
integrated copper lines.

Perlos liked it and wanted to continue development with
us. This could be the first big customer for our technology
- not a mobile phone company themselves, but a
component provider to many of the world's leading mobile
firms - including Nokia... Perlos could be the convincing
supplier we could never be as a small outfit. And they
would supply my tech!

At least ten meeting were held before anything really
happened. We dug into the technical details with Perlos to
such an extent that I became much more masterful in
plastics tech during that time. It was all I could really hope
for such a co-operation. In a steady Finnish style, no
hugging at the beginning of the meeting, no unnecessary
small talk, no bragging about solved problems. Just head
on into the problematic parts of the equation, and a calm
presentation of opinions and options.

Actually, I felt I could work for Perlos. It was a strange
feeling; I earlier thought I could never feel for a big
company. But something was so right in there. I even liked
their CEO, Ismo Rautiainen. A big surprise that I really

appreciated was that their CFO Tage Johansson participated many of the design meetings and gave his financial estimations of the tech's profitability throughout. This was really "design for minimal cost" - and I liked that too.

Maybe to a small company artist like me it is sometimes calming to be among big company processes, three-year plans and the 'real professionals'...

Not surprisingly the only thing lacking from the co-operation was a major transfer of funds from Perlos to F-Origin. They considered this phase more of a marketing effort from us and pre-planning from them (of course...). But based on the mutual enthusiasm I was sure this co-op would lead to financial gains for both parties, and I continued the prototyping and the meetings. Actually, this was one of the very few business relationships I handled almost completely alone.

Meanwhile, the real giants had started to move in our favour...

29.
Apple round 2

Whilst I was wheeling and dealing with Perlos and PlastiTouch, our American sales master Gloria forwarded us very interesting message from Apple.

"We would be very interested to see a prototype that would demonstrate your touch screen technology functionality on a user interface similar to our iPod touch wheel". So, they wanted to see if the sliding finger-touch they used in their touch donuts (remember how iPods worked back then) would operate with HaptiTouch. No problem!

We - F-Origin - subcontracted my original System Software team from the already severely limping MyOrigo to do the actual coding work. Those guys had already established their own small company so they were actually independent of MyOrigo anyway. Their firm offCode Ltd took the job and my trusted long-time SW guys did the work. It was to be a relatively simple task; after all, our HaptiTouch was much more sophisticated compared to what Apple had at the time, so we'd have no major problems emulating the function of their quite primitive capacitive touch system.

TECHNICAL & IRRELEVANT

The only challenging issue was related to the physical nature of HaptiTouch. As HaptiTouch -

with touch recognition based on the physical distribution of touch forces or weights - does and cannot detect touch with zero force, the user has to apply some miniscule force with his finger before the touch is detected. In any static touch operation this is never a problem - after all, buttons are always pressed, but in finger slide operations it was sometimes tricky to find the right combination of sensitivity and reliability. In this case, the minimum touch force to rotate the donut with sliding touch was selected to be 50 grams (0.5 newtons), and the touch force to perform a "click" was 300 grams (3 newtons). This provided an effortless operation that matched the existing iPod operating feel.

Related to the same non-zero-force issue was the physical friction of the finger slide on the current surface of the myDevice touch screen. The screen material and finish was certainly the exact kind of transparent plastic (Polymethyl methacrylate PMMA) that provided the best possible transparency and durability against scratching. The finish of the surface was very smooth; it did not yet contain those designed micro-imperfections that would later become essential to the modern touch screen to enable an effortless non-sticky feel for finger slide. Remember - practically the only slide operation we used was the myBook page flick left-right and vice versa, so we didn't concentrate much on how the sliding felt. For us the optical quality (and development time & cost...) of the plastic touch panel was much more important. So in practice, setting the sliding force also had to be adjusted and filtered in such a way that the inevitable momentary finger sticking would not confuse the algorithm in detection of a slide along the donut circular path. Not surprisingly, after finding the initial touch forces to start the testing, the "allow finger drag & momentary sticking" algorithm robustness was the hardest part of the development.

And so the System SW guys created a neat software application for myDevice that turned the screen into "an iPod", the emulation of an iPod, with its touch donut in place on screen. And the donut operated just like on the original iPod, sliding a finger on its surface and clicking it provided those same familiar operations. A couple of these prototypes were demonstrated to Apple.

The next message that came through to us from Apple was:

"Based on the prototype demonstrated to us, we are interested in utilizing your touch screen technology (HaptiTouch) in our next generation of iPods. We'd like to have your myDevice phones and software development kit for further study".

I clearly understood what this would mean, Apple would get to see and understand the workings and the internal secrets of our myDevice and its software architecture, including getting a sneak peek (via SDK) of the virtual screen and myBook swipe -systems.

The mind-set of our American sales team was that it was now or never. This was a real chance to do business with Apple, and any 'unnecessary' secrecy would only result in a drop in their then high interest in us. So we should send all the info we could - and also, send the myDevices and SDKs to study!

I agreed. After all, all the confidentiality and hiding the secrets of our tech for half a decade had not resulted in any business, just an endless line of expensive non-disclosure agreements that never meant any money for anybody but the lawyers. I was tired of hiding the facts and ingenious implementations of my inventions.

And so; the keys to the UI kingdom were hand-delivered to Apple.

30.
MyOrigo bankrupted – hair partly gone from my head

While I worked on with Perlos and PlastiTouch, guided the Apple "iPod emulator" prototyping and tried to organize the F-Origin IPR and documentation portfolio into a more convincing set, MyOrigo went bankrupt due to lack of funds and prospective customers in September 2004. All the remaining employees were let go, all the ongoing operations completely terminated. The office was immediately closed.

I saw my original dream - a physical device based on my own innovations on the global shop shelves - crumble into ruins. There would be no MyOrigo myDevice that the People would know and buy - ever.

My hopes of getting something out of the mess were now totally based on F-Origin. Teemu told me he had already been sure a year ago that MyOrigo could not make it, knowing the problems of the ODM approach at the high end. At a logical level I had accepted that too, but emotionally not... up until the very end I had still hoped MyOrigo would survive and myDevice would reach the consumer markets.

Well, it didn't. And that's that. In three days following the bankruptcy, I lost a palm-sized patch of hair from the back of my head. The rest of my hair didn't turn grey overnight, so it seemed I was not done yet, this was just an early warning...

A handful of the best and most trusted patrons from MyOrigo were hired by F-Origin. Markku Virta as Program Director, which in practice would mean the leadership of any practical project we would (hopefully) get from big clients. David then would have the freedom to be the hands-on technical project manager as Markku would handle the bureaucratic fish pool. Markku had been the display and camera sourcing and measurement specialist at MyOrigo, and hence had already quite extensive connections in the industry, from the "kitchen door" of component vendors. He was the spitting image of any classic Hollywood movie office clerk. A smooth and calm exterior combined with an analytical and peacefully spoken voice were just what international customers would need to be convinced of our professionalism. He was usually by the books, meaning, following design and good practice guidelines to the letter. Certainly not the most innovative guy in the world but I was there to handle that part. Markku would balance my often customer-scaring radical approach as the necessary non-intrusive counter-weight.

However, in my mind the most important recruit was Juha Rytky. A bald Mafioso-looking chairman type, almost exactly my age, had been cast as the MyOrigo Sourcing Manager back in the spring of 2001. Juha had back then an already-impressive background in the industry - he'd been Polar Electronics' (the world's largest sports heart rate monitor manufacturer) Sourcing Manager for some time, and wanted to jump into a smaller company to try out his wings and climb up the company ladder faster. Well, he certainly did, since at the end of

2003 he was already the COO of MyOrigo, handling many aspects of the daily business in addition to the sourcing process.

Sourcing Manager, wait a minute... has something to do with buying? Correct, it means the guy who ultimately manages which components we were going to buy and from whom, to build our mass production phone. Since a mobile phone has hundreds of separate parts - components - in its BOM (Bill of Materials) and at least thirty different vendors (companies producing those parts), the sourcing manager has to negotiate the best prices he can get for all those chips and bits, with all those companies. And he has to confirm second - and third - sources, if for some reason the primary supplier were unable to provide the components in-line with an agreed price or schedule. All that has to go hand-in-hand with the actual product design process, production schedule, deliveries to the factory, and so on. And since for small companies almost everything had to be JIT (just-in-time) to keep the cash flow and funds tied to to-be-needed-next-month components bearable, the schedule management had to be top shelf.

Juha handled all that and more. That was the good side of his personality: the ability to synchronize the motion of very many moving parts - whether it be component deliveries or customers' deals - with ease and never missing a beat. It was impressive, really, for such an unorganized fellow like myself. The bad side? Well, like so many talented guys, he was quite easily irritated by idiots and underperformers, and could resort even to physical violence in dealing with such people. That combined with alcohol usage habits almost similar to mine meant quite an explosion-prone character under the professional shell.

I thought that kind of behaviour was to be expected from

Juha because of his bright skillset (there is always flipside to the coin), and supported him always. I liked him! In a way it was easy because our competences and ideas never overlapped much, so there was no competition in the air between us.

On a side note, Juha had become my brother-in-law during 2004. My sister Sini was working for MyOrigo doing marketing, and I introduced Juha to her, even invited them together in a dinner one night which supposedly kick-started their romantic relationship. It didn't bother me at all, knowing both people I thought their relationship might be a workable deal. So they married and had kids a couple of years later, after a bit of turbulence in their early mutual life - as to be expected from such strong personalities (my sister was not totally unlike me in her persistence and sometimes stubborn behaviour).

Juha stepped into F-Origin like it was a family business, then. He became the COO of F-Origin, with a wide role basically overlapping with Teemu's CEO role whenever necessary. Utilizing the humongous amount of connections he had from the sourcing side, he was a base player of our team.

So the team was now more complete. J-P, Teemu, Juha, David, Markku, me, and a handful of other junior managers mainly from MyOrigo.

In many aspects, F-Origin was the "New MyOrigo with lessons learned" for me. To keep sane, I had to put all my efforts and future on it, forget the fall of MyOrigo, and just keep pushing with Perlos, Apple, and hopefully others to come. I had to learn to take the criticism and even flat out hostility when walking around in the small city of Oulu - some of the ex-MyOrigo people felt I had let them down, let them go bankrupt, had 'stolen' the crown jewels (IPR) from MyOrigo to my new company which they had no part in. Luckily I never depended much on other people's opinions, so this was no major stretch for me, actually. Inhuman? Sociopathic? No, just the necessary self-protection element for any radical inventor.

The press also woke up to the MyOrigo crash. Earlier, whenever the media was interested to do an article about us or our revolutionary myDevice, there was always an eager CEO or Sales Manager to take the credit for our inventions and ideas. Now, I finally got the press space I had been deprived by those glory-hungry fair wind sailors. However, not in the best of flavours: "The hype ran out of oxygen", was the tone of the press stories now.

I stopped reading local newspapers and marched on, towards the sure-to-come glory days of F-Origin.

31.
Biggest deal ever –
myDevice 2 for Samsung!

Every fisherman at least subconsciously wants the Moby Dick, but not everybody gets to hunt it down. We had been fishing for it for five years.

Now we had it on our hook, in the end of the year 2004. Samsung - that South-Korean multi-industrial mega-mammoth making everything from tooth paste to mobile phones to cars - had taken the bait, studied our IP and technologies thoroughly, and made the decision:

F-Origin was going to build **myDevice 2** for Samsung!

The negotiations had already started at the end of 2004, led by Teemu himself. What started as a simple idea of integrating our HaptiTouch or Virtual Mirror/virtual screen tech (newly coined "Iris" by me for F-Origin marketing purposes) had expanded into a full phone design - just what I had been hoping for at MyOrigo! Although it was not yet clear what kind of compromise of features the final phone resulting from this co-operation would be.

I had embarked upon a series of very long-haul flights from Oulu to Seoul in December 2004 to seal the deal.

David was also there, as was Antti Ylönen, our dependable and open-to-new-ideas HW design chief from the MyOrigo days. The route Oulu-Helsinki-Frankfurt-Seoul gave me a lot of time to ponder the exciting situation. I remembered well what Teemu had told me a week previously about Samsung's company policy - they tend to order and start ten projects from various subcontractors and partners, pay three (partly), and finalize only one of those ten. But as an inventor, I had to be sure that it would my project that would be carried to the end - to mass production by Samsung. This face-to-face meeting session with Samsung's Senior VPs and their design team was the necessary prelude to convince them to start the project full steam.

My intention was clear. Since it seemed I had been offered this chance; I wanted to remake **myDevice**, do a revamped version of it, with a new component and display tech available in 2004. Thinner, better, faster, and even more revolutionary than the original one. All the key tech of the original myDevice would be there, of course. HaptiTouch, Iris (Virtual Mirror + virtual screen by MyOrigo lingua), myBook with swipe, full web browsing, everything. And with Samsung muscle, it could be a feature-rich phone too - all that stupid MMS/Video call/Bluetooth -stuff could be built in.

For me, this was much more real than the still very vague possible co-op with Apple, or even the PlastiTouch project with Perlos. I was fond of PlastiTouch of course, but it would take a long time before I could see it really used in an actual mass product. On the contrary, the Samsung myDevice 2 could be on store shelves within 12-14 months, if we danced the samba just right with Samsung.

The Lufthansa plane landed in Seoul Inchon International Airport. My brain had been pumped from 1 ATM to 0.7

ATM and back six times in the last 32 hours, so I felt like a zombie on weed. With David and Antti, we took a taxi-ride through chaotic but strangely American-looking Seoul and went to sleep in a hotel located near Samsung's headquarters.

Next morning the endless stream of Samsung-style meetings started. It was more like an interrogation than a bilateral meeting. The Samsung guys were trying to get us, find any cracks in our story or in our tech. I imagine that is their form of NIH - the high bosses had said they wanted to try this with us, but almost all the guys below didn't want to do it. Why would they, it would be like admitting we were better at their jobs than them.

When one team of meeting partners - whether it be mechanics, industrial design or software team - wore out, another one stepped in to start roasting us. We were naturally given no rest. In between these meetings some VPs and Senior VPs came to meet & greet us with extremely positive comments - they were the good cops. "Let's do this! Let's start the project and change the world together. Haptii---tatsii!" they fanfared. And after five minutes of this kind of handshaking harmony, back to the meeting room (roasting oven) where the next team of doctors of technical sciences and senior analysts were waiting with their endless trick questions.

I grew to appreciate Teemu more and more during that trip. He acted in the only sensible way in this situation and didn't really miss a beat. Any time the opposing party got too hostile (tried to push the limits on the interrogation too far), Teemu slowly got red from the neck up, like mercury rising in a thermometer. When the 'level' reached his hairline things started to happen. A lot of things and fast. He stood up, started to shout, still smiling elegantly, stating that he has done this so many times that he knows how this is done and now they - Samsung - are doing it wrong. Always after such burst there was a small pause in the meeting, and then it went on with a more civilized tone for a while. Teemu was the reset button of these almost intolerable sessions...

The evening of the first extremely tiresome day Teemu took us to see the wonders of Seoul. He had visited here already a couple of times, this time arriving a day before us, and knew the stunts needed to have decent time here. He explained that South-Korean culture is a weird mix of American, Russian and Chinese cultures and habits. Roads and cars are pretty American in style. Food is a mixture of Chinese and Russian, also Mongolian cuisines. And people, they value entrepreneurship but are still artfully cunning like Russian businessmen.

Our first visit of that long evening was to the Korean Grill restaurant that had a (new to me) table grill system where everybody basically cooks their own food. Raw ingredients, meat and fish, were brought to the table and then you just grilled your own! It was unfamiliar but tasty and fun. Nowadays you can easily get that kind of experience of table grilling in every major western city, but back in 2004 it was unheard of in Finland at least.

The local rice beer and heavier sake-type liqueur Soju flowed. The moods hammered by long day of sheer insult were toned down, toned down slowly to better and more optimistic ones. We would strike a deal with Samsung, and smile all the way to the bank after all this bullshit had been tolerated. After all - Finland is the only continental country in Europe not occupied at any stage of the Second World War - not even by Stalin - so we could take it all, easily. Or so I thought, after my 7th glass of Soju and half a kilogram of medium rare grilled meat.

The evening progressed into the typical "western businessmen in foreign megacity" type of a deal, which Harri had so clearly explained and demonstrated to me years back in Prague. Korea had a surprisingly vast number of so-called Business Clubs built into its major cities - Seoul had one on every city block, no matter which direction you walked. They were hootchie-houses of course but kind of a soft porn kind, with the focus on drinking and lap-sitting, singing karaoke with the local bimbos, and so on.

What later followed in that evening will stick in my limbic memory. We - the team of four Finnish guys - took on the task of finding a really big-titted escort in this city where most females didn't have any voluptuous features at all, let alone in the upper body. This was my preference, of course, others would have been happy with what was readily available. And this would be more difficult than finding a needle in a haystack.

Teemu was the man for the job: Expressionlessly, he took a five centimetre stack of 50 Euro bills in his left hand, and started to ask every single person he came in contact with in the nightclub we were visiting: "Do you know where we could find a really big-titted but thin girl?" and pushed a fifty in the questioned hand. The stack grew thinner but we had no solid results. Teemu took another approach: he took a 2-minute lesson from a bartender (paying a 50 for that too) for how to ask the same in Korean. It sounded something like "우리는 큰 가슴의 여자를 찾고 싶어요. 당신은 도와 드릴까요? ". And so the crusade continued, with Teemu asking this over and over again, handing out 50s left and right. It was clear the Koreans thought somebody was giving away money for free, and the lines started to form. Teemu decided to re-evaluate his strategy once more and looked up with some frustration in his body language.

And just that moment, a big-titted, slim-bodied Korean Earth Angel walked down the nightclub VIP stairs. She was already waving to us. Teemu saw her and waved back, pointed to me making sure the Angel saw that and made a small bow towards myself. Angel locked her tea plate - sized greenish eyes on me and walked towards us catwalk-style... It later became evident that the nightclub owner in situ had heard about out crazy goal, and decided to humour us with the best he'd got. And so he did - after all, it was not the worst marketing for a dime in a dozen nightclub that businessmen were freely handing out 50-euro bills...

Thus the night was saved.

The rest of Seoul trip was more of the same interrogation/complimenting - the good cop/bad cop - stuff. We got through it reasonably well, the Koreans seemed to be somewhat satisfied because it was obvious we knew technically what we were talking about. I got to fly home to spend New Year with my family. I was back before Christmas Eve.

32.
Samsung –
Level 2

The year 2005 started with high hopes. I had forgotten the MyOrigo bankruptcy misery - now I viewed it as a necessary middle step towards real success, like a learning process. F-Origin was much cooler and actually fun to work at for a change, to no small avail the effect of generally Johannes-compatible colleagues.

I got the F-Origin IP sorted and straight - we had the 6th largest motion control & touch screen patent portfolio in the world! And that's not bad for a company of 10 guys... But like I said before, patents mean nothing without infringement from a big party. And our patents protected ourselves (tried to at least); they were not designed to attack.

So real business would have to do it for us, and that's the way I like it. Samsung was almost ready to sign a multi-million Euro deal with us, based on the positive Far East meetings in December.

Teemu worked in the background with the actual fiscal agreement with Samsung, which was going to be a multi-layered, multi-milestoned mammoth of a deal. Meanwhile in Finland I was already gathering my groups to start the design of **myDevice 2** for our South-Korean Moby Dick. I was extremely excited; this was sort of a new type of enjoyment for me. I got to improve my original myDevice design, having not really to invent anything new, just organize original innovations into better and clearer function and sleeker style. It was like admiring your previous work and fine-tuning it here and there, like a master painter making the final adjustments to his almost-ready painting after a month of pastis-filled holiday in Paris.

I needed a mechanics guy that could improve the HaptiTouch based on my recent findings with PlastiTouch prototyping. We could not yet directly utilize PlastiTouch in this Samsung project as the technology would be production-ready only years later - if Perlos pulled through and did an agreement with us. I found Erkki Hinkola, an ingenious but calm and listening designer who would design the fiddly and sometimes unreliable HaptiTouch of MyOrigo into HaptiTouch 2.0, more sensitive and robust, but cheaper-to-produce version of it.

Hardware - the electronics of the device - were really not a problem since Antti was already working on it. The basic HW architecture would be very similar to the original myDevice - the earlier design choices were solid and good, no need to change what we knew was working well. We would need to outsource the antennae and RF (radio frequencies) design but I knew where to find those kind of guys in Oulu - it was one of the mobile design centres of the world back then.

I knew I'd got system SW covered with those ex-MyOrigo offCode friends, and myDevice general architecture would do just fine especially as David was here at F-Origin to mod and adjust as needs arose. But I needed an application software team, since there would be a ton of new features and applications compared to the streamlined (meaning minimal, due to lack of development funds) app selection of MyOrigo's original myDevice.

I decided to pay a visit to my old employer before MyOrigo was even started, CCC Software Professionals. They had a 700-strong professional SW team and quality control methods I knew and liked. So, I would let them do the 'ditch digging', the humongous amount of generally uninteresting application software work to fully pimp up myDevice 2. They had done many projects with Nokia so they knew how the old-fashioned mobile phone SW was done. And since I knew they were familiar with Java too, I could pretty easily teach them our current (the most modern in the world, of course) Java-based application architecture.

Teemu and I did a presentation to CCC in the early spring of 2005. Since I left CCC on good terms, not burning any bridges back in 1999, we were quite easily able to find an agreement that suited both parties. It would be expensive to us, but since we had "open books" cost coverage promised by Samsung, it wouldn't really matter - and I knew these guys would get the job done even with my liberal spec. On the CCC side two of the negotiating parties were actually my old bosses, and they remembered that technically I was solid. Whether I was solid businesswise, would soon be found out.

The final thing CCC insisted was that they could invoice the bulk of the work directly to Samsung, to get to list those guys officially as their direct customer. OK for us, actually better because our financial exposure would be likewise smaller. Samsung also accepted the arrangement - after CCC made their own sales pitch to Samsung VPs in our presence. CCC had much nicer headquarters than we did, so the meeting was held there...

And so, with these developmental elements in place, the deal with Samsung got done. CCC had their own agreement, pretty similar to ours. I thought it was a sweet deal, covering our and CCC's costs fully if we did it smart, and then paying royalties of several dollars per produced device to F-Origin. We would get rich, finally!

The most important thing to remember was that all the payments to us depended on the specified milestones, "checkpoint Charlies", and if we did not reach them exactly at the promised time Samsung would have the right to cancel the project. The payment terms if such a cancellation occurred were convoluted, and I decided not to let them ruin my day, since my project would not be cancelled. Let Teemu and Juha worry about that!

Samsung decided to handle the project through their European resources based in London. And the leader of this project on their side would be... Torde from Sweden, our neighbouring country! Torde had spent most of his career at Sony-Ericsson so he knew what Scandinavian mobile phone design meant in practice. We thought it would be easier to co-operate with him than some South-Korean, poor English-speaking manager, so good so far.

Everything was set. It was March 2005 and I pressed the metaphysical start-button with Teemu. F-Origin running costs increased overnight from couple of 10k to more than 150.000 Euros per month. The hardware and software design process started immediately at full steam, even when the final features were not fixed yet with Samsung – the most important things we knew and were able to predict.

My next task: specification meeting at Samsung European

Headquarters in London. Their spec team would fly there from Seoul, and the relevant Samsung UK employers would also participate.

I knew the spec would be international and localizable, so there was bound to be a lot of talk about different language versions and locales. Luckily, one of my oldest friends from childhood, Kimmo Myllyoja had worked with that specific area (localization) at Microcell - recruited there by me even before MyOrigo started. So I subcontracted Kimmo who was already running his own company, he had resigned from Microcell some time ago to set up his own lifestyle. Kimmo and I flew to London with noble hopes of very productive meetings. The pow-wows were supposed to last two days, three if necessary. And after that I could fly home with a full, approved user interface and application spec in my thirty Euro briefcase.

Little did I know it would be a set-up.

33.
The London trap

It was a massacre. A carefully planned, well executed system to gather proof that we - I - would not be qualified to run this project. It was designed by the middle-managers who did *not* want to work with us, who wanted to design their own phones and not let "guys from the woods" do it for them.

To understand the extent of this, I need to explain how that particular series of meetings *should* have gone to be productive.

In addition to being an inventor, I was *the* user interface and applications expert, globally top shelf by any standard. I would have presented the guidelines we were going to implement our myBook, HaptiTouch and Iris to suit Samsung's style and image, and they would have commented them, on the general level first, then going into some detail of the critical UI parts like the menu pages' arrangement and so on. Then, on second day probably, we would have looked at an overview of the apps that would be factory installed to the device, and the general spec of those apps in a wide ballpark. The detailed implementation of everything would follow pretty much from the GSM and other standards for many applications, and the general myBook user interface ideology would surround all of it. Clear and simple to discuss. And I had the materials for it, as well as many MyOrigo's myDevice phones to try out the previous version of our UI and apps. Then Kimmo would follow with his world-leading localization expertise, he knew how to implement and spec far eastern, Cyrillic and even Arabic languages on the device. So the localization spec could be overviewed in great detail.

So that's how it should have been, in a rational world. But it wasn't.

It was a trap set up in the most offensive way they could have figured out in South-Korea. There was a Specification Manager, a young girl who had to be less than 30 years old, who could have been brought there directly from any Victorian chamber of horrors. I suspect she was not entirely human. It's hard to describe her looks other than thin, very Korean and extremely devious. She sat at the central position on the other side of the table, surrounded by black-suited bodyguard-type Korean fellows who were all managers of this or that, by their cards. Kimmo and I sat on the other side of the long table, with the Samsung western employees and Torde.

Then the show started.

The girl had three extremely thick black books on the table. She opened the first one, and started with monotonous, but strangely annoying voice:

- A5/1/2/3/8X. Do you comply?
Me: (Ahmm, you mean the GSM encryption algos? Sure, it's built-in to any modem today, but what does...)

- AGCH. Do you comply?
(Errrh... if I remember correctly, this is related to access the granted channel or something. I think we do, but hey,

I think this is unproduc...)

- CCCH. Do you comply?
(OK now, stop, this is not what I came here to do. I can't remember what the acronym means and...)

Guy from the left side of the girl: "Tut tut tut (shakes his head)". All the other black suited guys also shook their heads and one of them made a note in his notebook.

- CLID. Do you comply?
(Now Ms Spec Manager, this is getting ridiculous. Do you expect me to remem..)

Guy from the left side of the girl: "Tut tut tut (shakes his head)".

And so on.

I kid you not: this went on for hours with tiny interruptions by the Samsung managers stating: "We need to go through this or otherwise we have to go back to Korea and cancel the project because of the lack of info from your side". And I was depressed enough to sit there. On a lunch break one South-Korean manager asked me: "How do you think the meeting is going?" I said colourlessly "it's unproductive and we should change the protocol" and continued to eat my tasteless British mush or something. He walked away with a narrow smile.

Kimmo sat there that whole day without really saying a word. In the evening he said: "Josse, I'm not staying here for tomorrow, this is a bad joke and extremely painful to watch. See you later, bye." And he flew back to Finland.

I phoned to Teemu and told him I was coming home, fuck

Samsung. He had an argument ready: "I know it is bad, it is the NIH internal resistance. But you can take it. Stay there for two more days and I'll give you... a 50%+ raise in your salary and all the flights from now on business class, out of my own pocket". I swore a bit and stayed. It was my future on the line there and then, we could not afford to lose Samsung. If I gave those torturers a reason to go back without finishing the meetings - because of me - Samsung would cancel the project and it would be over.

The most humiliating thing was not what they were doing to me. It was what I did to myself. I listened to two days of shear insults and mockery in the hope of continuing the project with these 'South-Korean friends'. I should have left after the first hour. I didn't. I had lost the edge, I was afraid. I got the fear of losing.

34.
Teemu and the explosive meeting

My previous London meeting had been more than a month ago and I actively worked to forget it. Those 'managers' went back to Seoul and the project continued, if not in high spirits, but anyway with high hopes on our side. Samsung had even paid our first quite massive invoice, which meant all was real and legit on their side too.

Now there would be a general project meeting where many mechanical construction details of the device would be agreed upon. Samsung had the best mobile phone mechanics in the world, even better than Nokia, so we knew this was important. Teemu would participate with me, as well as Paavo Repo, our new project manager. Paavo had taken over since Markku left couple months prior to a start-up high-tech company - and I thought Paavo would be even more suitable for this project, since he would be able to push back when Samsung played their tricks. Paavo was a 50-year old seasoned program manager veteran, had led mobile phone design project that resulted in millions of produced phones - on schedule - and he was also a wrestling champion! At least he could bend those Samsung managers into a human knot if the need arose...

Samsung had brought to the scene the big bosses this time, no middle managers. It seemed they started to take our co-operation more seriously. A mechanical guru who was also a production factory boss from Samsung, was on the scene to lead the discussion about HaptiTouch.

We called the guy Fred Flintstone not only because of his looks, but because his robust handling of our prototypes seemed to concur with that image. He repeatedly hit his own forehead with our MyOrigo myDevices we had brought to demonstrate the benefits of HaptiTouch (the Koreans pronounced it Haptii-tatsii? with a clear question mark on the end every time they said it) in practice. Maybe he got a better feel on his forehead than with his hands, or maybe he needed a hard enough surface to hit our phone against to get a handle on how rigid and robust the construction was.

He spoke only few words in English, he had obviously got his position via hard work and dedication through many years of battle. He sure wasn't royalty so that was not it... I kind of liked him, he was a practical guy I could talk to. If only he could understand me.

The other Samsung bosses were mainly annoying. One of them kept repeating that all the Samsung phones had to be able to take a fall from 1.8 metres onto a hard surface, and he didn't believe our tech would survive that. He kept nagging about it and I saw that Teemu's neck started to get red from the bottom up. I knew what it meant and whispered to Paavo: "Be prepared for some wrestling moves soon, something is going to happen and those hard-core Korean bosses, especially Fred, might actually do something physical about it". Paavo visibly stretched himself a little...

It was on now. Teemu's face was all red, only the forehead remaining greyish white as a sign of remaining sanity. Without saying a word, he snatched the latest, most expensive Samsung 'smart phone' from the hand of one of the Samsung managers. Then he hopped onto the meeting room table, lifted the Samsung high above his head, and dropped it. It hit the table full force and parts sprayed all over the room, the display shattered. Teemu stated bluntly: "See, yours couldn't take it! You were not telling the truth!"

Time stopped, physically. Nobody moved. Teemu was smiling now. Paavo tightened his grip on the seat handles, obviously ready to get up in a fraction of a second if things escalated. We could only imagine what kind of insult this had been to the South-Korean bosses, maybe they lost face or... something with this stunt.

10 seconds passed. 20. Teemu quietly leaned towards me and I gave him a myDevice. He dropped it from the same height as the now-dead Samsung phone. It just bounced off the table, the battery separated from the device, but Teemu put the battery back and made a phone call to me and said triumphantly: "Seems to be working!"

As if in common agreement everybody sat down, and the meeting continued without further drama. Even Fred accepted that Haptii-Tatsiii? would be a part of the future Samsung myDevice.

Just a week after that meeting, the Samsung guys - some Korean small-time managers and the whole of Torde's UK group - visited us in Oulu with Torde leading the pack. We got them very drunk in the best restaurant in Oulu (called Üleaborg), one of the Korean guys passed out in a near hospitalization-condition. Torde was Swedishly happy when I helped him out of our minivan at their Hotel. I unintentionally mopped a part of that hotel's foyer with Torde which I am sure he forgives since we all had fun.

Later I heard that that Korean guy had actually became colour blind during the visit. Truth or not, but I know that the Finnish shot "Marskin Ryyppy" can have all kinds of interesting side effects - it also contains absinth. The restaurant bill showed we (10 of us in total) drank 196 of them during the evening, after we had had several bottles of fine wine during dinner. Since there were three non-drinkers, that means more than 25 shots for each of us. Welcome to Finland!

The project continued and summer was here. I was high again – for the moment...

35.
The Fall

I call years 2001 and 2002 "the golden years". Those were the times I thought sky was the limit, and my childish belief in conquer-all triumph was at its peak. From 2003, it was much rougher at MyOrigo. F-Origin made me believe that the saying no pain, no gain held true. The pain of MyOrigo's bankruptcy would lead to the final gain, the Samsung myDevice with each produced unit hammering a profit for our company. It was to be the true iPhone, and Samsung had a chance to get it back in 2005!

The bright Finnish summer was at its peak. We had the first functional prototypes of myDevice 2 in our hands, with the system SW and bootloader running. The protos were exploded PCB types, the final components with HaptiTouch mechanics had been placed on a 20 x 20 cm circuit board to make measuring and debugging of each individual component easier. It worked like a clock with its brand new Intel Bulverde mobile processor, about four times more powerful than the original myDevice's Intel StrongArm.

We had a milestone coming up. Samsung was sending some engineers to check that we had actually reached the points specified in our agreement. They would come on the weekend, the idea being to see that we were working day and night to make this happen.

Sometimes bad things just clutter up. The smallest details can turn the tables, as I have seen throughout my life. The devil is in the details.

The first thing to go wrong was the Perlos co-operation - the PlastiTouch futuristic deal. When we were busy with Samsung, Teemu had sent his lawyer friend Pekka to negotiate a monthly retainer fee from Perlos to F-Origin to keep the project going, even before the final licensing agreement was done. Perlos had offered 30.000 Euros per month, which I thought was extremely generous. For reasons still unknown to me, Pekka had physically ripped the papers in half when he read the offer, and walked out of Perlos without saying anything. Perlos still wanted to continue even after that, but the relationship would have to be rebuilt, fixed by me. I didn't have the energy or the time to do that then, with Samsung pressing like a hydraulic injection moulding machine cylinder.

The second thing was the growing disagreement between Jyrki, Teijo (Jyrki's partner in many of his tech-companies) and Teemu. It was obvious there was something unrelated to MyOrigo in the background, but I never stuck my nose in. Let adults deal with their differences in their own style, I thought. But this tension was affecting how Jyrki behaved towards F-Origin and - also me.

Misfortunes never come singly, or as we say in Finland, no second without the third.

The third came. Maybe not knowing about the Samsung visit, Teijo decided to lock us out of our own office and laboratories on the day of the visit. Paavo took the Samsung guys to our office only to find the doors were locked from the inside and he could not get in. The

Samsung guys stood there for five minutes while Paavo tried to resolve the situation by telephone, then they left quietly and never came back.

Samsung cut all communication with us, immediately. They did the same towards CCC. Even Torde was hard to get on the line and he couldn't say anything.

We had reached the status needed to get payment for a second milestone but Samsung never got to see it. They never paid a dime to us after that incident. We owed money left and right, our monthly costs had been 150k+ because of the heavy subcontracting and super-fast project build-up.

Apple had received all the SDK materials and protos, but was silent. We didn't know what they were doing, the only message we got via our American contacts was that they need more time.

With Samsung and Perlos gone - and Apple on hold - we had nothing.

My dream was over. F-Origin filed for bankruptcy in August 2005, leaving more than a million Euros of unpaid debts to many companies - many friends - behind.

We met with Teemu, Juha and Paavo for a final get-together on the terrace of the restaurant where we got the Samsung guys stoned just a couple months before. Teemu took his mobile from his pocket - it was Samsung's latest and greatest model which he had been using to get to know the Samsung style of design. He threw it into the sea which was just metres from the beautiful terrace.

I watched the sea glimmer as it swallowed the Samsung phone.

—

36.
Catharsis –
iBall, Nintendo Wii and
Déjà Vu

After the fall of MyOrigo and F-Origin, my inventor's instinct was still clear and strong. I was 30 years old and at the top of my game – just temporarily in dire straits. The financial failure of the origo-companies led me to believe that if one wants to bring new hardware to the world with limited funding, a small and cheap device is the best approach.

During the origo-years my knowledge of practical consumer level motion sensor technologies had reached heights that only few could match. There were numerous studies on the subject done around the world, and everybody in the business was generally excited about the possibilities of motion controlled user interfaces. But end user-level applications and devices were utterly lacking. Garage prototypes and university research programs were created, doctorates granted, the internet buzzed – but no real motion control device was available to the general public.

It was obvious that I was in pole position to create something real, mass-producible and cheap that would utilize accelerometers, the main motion sensor type used in the myDevice phones. I had toyed with the idea of a trackball mouse without a physical cage for some time, even when F-Origin was still going strong. Now I was free and almost penniless, so the time seemed ripe to actually build the magic ball. That would be just what the world needed to spark the motion control revolution, making old-fashioned mice and trackballs obsolete! I needed it too, to get my inventor act together with new investors, new money and, hopefully, sales that would finally make me rich...

TECHNICAL & IRRELEVANT

Accelerometers alone can only statically determine the direction of the gravity vector, so in theory they can only measure the device's orientation around two axes. The orientation about the gravity vector (vertical from the ground) was unknown, and my free-air trackball would have data from just those two axes. Luckily, just around this time in 2005, cheap 3-axis magnetometers became available. I figured out that the magnetic field vector and gravity vector together would result in a full three-dimensional knowledge of the ball's orientation in the real world, and thus enable the functionality of the free-air trackball.

Why were all the axes needed anyway? After all, a trackball is only rotated around x- and z-axes, not the vertical y-axis. So could an accelerometer alone do the job, just by following the gravity vector? Well, no. When the ball is rotated over and over again, it naturally loses its original polarity, the understanding of where is forward and back, left and right, without any external reference. So after a while, when rotating the free-air trackball to the right from the user's point of view, the ball might think it is rotated forwards – or to left. The gravity vector could never provide a critical piece of information, the user direction reference, that would be needed for any practical usage over a couple of seconds.

```
So, I quickly designed the basic architecture
in my head, comprising a 3-axis accelerometer,
3-axis magnetometer and an air pressure sensor
to measure the "squeeze" of the ball. Squeeze
detection was needed to press the mouse button
- since there were not and could not be any
external buttons, the whole ball surface had to
be made sensitive to touch.
```

The 'Free-air trackball' idea was interesting enough to convince my long-term colleague and COO of the origo-companies, Juha Rytky, to join my efforts in building and marketing the invention to the real world, the consumers.

Together with Juha and a couple of other friends, Tommi Korhonen and Rami Niiranen, we implemented the device at a prototype level in the autumn of 2005. By spring 2006, we had funding in place, our new company Ball-IT Ltd was up and running and a golf-ball sized revolutionary motion control device in our hands. We were able to convince one of the largest sensor manufacturers in the world, ST Microelectronics, to issue a press release about our invention, promoting ST sensors at the same time. This was published well before the introduction of Nintendo Wii.

The original text of the press release read: "Making its debut at ST's stand at Electronica 2006, the smart golf ball-sized object can operate as a free-hand personal computer mouse, compass, measuring tape, pedometer, or a 3D-object controller."

That same November 2006, the Nintendo Wii was launched featuring "motion control for the first time in history". They claimed it to be the first... this feeling would become a familiar one to me, one could even say a form of déjà-vu, just a few months later in January 2007...

Well, it's nice to have the big guys trying to do the same as you, I thought. So we progressed with our Ball-IT ball invention, naming the device initially the BluetoothBall. A year later we finally turned our attentions to the gaming markets when it became obvious that a new mouse type for PCs was not going to sell. I figured the reason for that was the same as why cars still have steering wheels.

Finally, I got to see my idea as a final product, branded "blobo" for consumers, on the shelves of shops around the world. More than fifty thousand units were sold. And behold, many customers seemed to like its simple, elegant approach to motion control with our cube-headed game figures and the "Fun & Fit" game bundle. There was even a TV Show in Finland created around gaming with blobo and it ran with good viewing figures for over six months.

Motion Control before Nintendo Wii

Through presenting blobo to potential partners, I also got to meet some interesting people in the industry in addition to all the mobile phone guys from my previous endeavours. Peter Vesterbacka, later known as Mr. Angry Birds, Henri Seaudoux, the headmaster and chief innovator of Parrot France, Alex St. John, inventor of DirectX and back then the CEO of WildTangent, Honkura-san, the head of Aichi Steel research and development and Mr. Tsu, the head of the China National Investment Bank. Interesting fellows and discussions.

The adventure with blobo was a lengthy one, ending after 7 years with the closure of the company in financial ruin. In many respects, it was a more exciting time for me than the origo-companies – I had more power, less friction and smaller team that I could handle and work with effectively.

The blobo-era was an amazing trip that took me around the world, finally culminating in 2011 with China, big partners and potentially big cash. Technically a great success, blobo (ironically renamed iBall for Asia!) failed spectacularly with the China venture capitalists and I was done for once again. But I've had my Catharsis; that is another story to be told in another book...

37.
Calvados –
January 2007

I was sitting in my modest home – a type of detached house built after the Second World War for the families of soldiers who fought on the battlefront – when Steve Jobs presented his latest innovation, Apple's revolutionary iPhone, on TV. The basis of the iPhone was clear from the first few minutes. It was an obvious copy, even flattering in its legacy, of MyOrigo's original device and ideas. Naturally technological advances from last half a decade had made the iPhone smaller, slimmer, and faster – but the foundation shone through like a pimple under powder.

I slowly turned my gaze to the original **myDevice** in my hand, it was still my everyday telephone. I hurled it at the wall full force, walking to my drinks cabinet. Took a hefty gulp of Calvados. Then another.

myDevice did not break – it is still operational, and waits forever to be touched again on the top of my bookshelf.

Epilogue
10 years later

Did we – the inventors, the engineers - destroy the future of society by corrupting the minds of young people, by giving world the idea of a modern smart phone, keeping them always hooked, tempted to stay "connected" by the smooth and appealing user interface?

Did MyOrigo and Apple open Pandora's box, bringing usability and instant connectivity to everybody's pocket, enabling 24/7 social networking? There is some comfort in the notion that it would have happened anyway, maybe later, maybe in a slightly different form. But it is done now, that's for sure. Facebook, Twitter, Instagram. Always looking for that next endorphin dose; next comment or thumb up to a recent selfie.

It is amazing how young people's ability to concentrate has deteriorated during recent years. In some countries they are changing house locks from four to three digit codes because youngsters simply can't remember a four-digit number any more without the help of their smart phone. They wouldn't be able to get into their own homes if the phone battery died.

It seems nobody can boil an egg, or write a one-page document, without checking their Instagram account in the middle.

What have we done?

The author of this book is currently working on revolutionary technologies that aim to free people from constant visual and aural overload.